"With his excellent new book *Ethical Responses to Nature's Call: Reticent Imperatives*, James Magrini has written an important and much-needed contribution to the field of eco-phenomenology. Not only does Magrini develop eco-phenomenology in an original way, but he also introduces a new stance on ethics and Object-Oriented Ontology. With his book, Magrini is clearing a path for us, living in an era approaching environmental catastrophe, to hear and listen to the call of Nature, that is, what Magrini names "ontological aurality." This inception of an original ethics of ontological aurality and excess can be compared to Martin Heidegger's insistence on "the silent power of the possible." In Magrini's version of Heidegger's adage, the environment, as an excessive object which reveals itself in its very withdrawal, is indeed the power of the possible, or as Magrini pointedly expresses it "a living force." This living force is the very possibility of a new foundation for an ethics developed out of, precisely, reticent imperatives – imperatives which, indeed, resolutely sound out of the book's call to our environmental conscience."

Elias Schwieler, Associate Professor of Education, Stockholm University, Sweden

"This is an important book that explores in a philosophically rigorous manner the topic of nature's 'voice' and its significance for achieving an authentic relationship with nature – that in turn is central to responding to the environmental crisis that currently besets us. Magrini has a gift for identifying the most germane issues and for interpreting and weaving together a wide range of cogent views. The result is a stimulating read that makes an imaginative and essential contribution to the literature in the field."

Michael Bonnett, University of Cambridge, UK

Ethical Responses to Nature's Call

Arguing for a renewed view of objects and nature, *Ethical Responses to Nature's Call* considers how it is possible to understand our ethical duties – in the form of ethical intuitionalism – to nature and the planet by listening to and releasing ourselves over to the call or address of nature.

Blending several strands of philosophical thought, such as Graham Harman's *Object-Oriented Ontology*, W. D. Ross's prima facie *ethics*, Alphonso Lingis's phenomenological ethics traceable to *The Imperative*, and Michael Bonnett's eco-philosophy, this book offers a unique rejoinder to the problems and issues that continue to haunt humans' relationship to nature. The origins of such problems and issues largely remain obscured from view due to the oppressive influence of the "Cultural Framework" which gives form and structure to the ways we understand, discourse on, and comport ourselves in relation to the natural world. Through understanding this "Cultural Framework" we also come to know the responses we continue to offer in answer to nature's call and address, and are then in a position to analyze and assess those responses in terms of their potential ethical weight. Such a phenomenon is made possible through the descriptive-and-interpretive method of eco-phenomenology.

This renewed vision of the *human-and-nature* provides direction for our interaction with and behavior toward nature in such a way that the ethical insight offers a diagnosis and provides a potentially compelling prescriptive for environmental ills.

James Magrini is Adjunct Professor of Philosophy and Ethics, College of Dupage, USA.

Routledge Explorations in Environmental Studies

Environmental Human Rights
A Political Theory Perspective
Edited by Markku Oksanen, Ashley Dodsworth and Selina O'Doherty

African Philosophy and Environmental Conservation
Jonathan O. Chimakonam

Domestic Environmental Labour
An Ecofeminist Perspective on Making Homes Greener
Carol Farbotko

Stranded Assets and the Environment
Risk, Resilience and Opportunity
Edited by Ben Caldecott

Society, Environment and Human Security in the Arctic Barents Region
Edited by Kamrul Hossain and Dorothée Cambou

Environmental Performance Auditing in the Public Sector
Enabling Sustainable Development
Awadhesh Prasad

Poetics of the Earth
Natural History and Human History
Augustin Berque

Environmental Humanities and the Uncanny
Ecoculture, Literature and Religion
Rod Giblett

Ethical Responses to Nature's Call
Reticent Imperatives
James Magrini

www.routledge.com/Routledge-Explorations-in-Environmental-Studies/book-series/REES

Ethical Responses to Nature's Call

Reticent Imperatives

James Magrini

Routledge
Taylor & Francis Group
LONDON AND NEW YORK

earthscan
from Routledge

First published 2019
by Routledge
2 Park Square, Milton Park, Abingdon, Oxon OX14 4RN

and by Routledge
605 Third Avenue, New York, NY 10017

First issued in paperback 2020

Routledge is an imprint of the Taylor & Francis Group, an informa business

British Library Cataloguing-in-Publication Data
A catalogue record for this book is available from the British Library

Library of Congress Cataloging-in-Publication Data
A catalog record has been requested for this book

ISBN 13: 978-0-367-72822-9 (pbk)
ISBN 13: 978-1-138-36789-0 (hbk)

Typeset in Goudy
by Wearset Ltd, Boldon, Tyne and Wear

To Laura, of course. You continue to bewitch me, much like *Nature* (*φῠ́σῐς*), in ways that defy language.

Contents

About the author

Dr. James Magrini teaches Western philosophy and ethics at College of Dupage and was awarded the 2013 Outstanding Part-Time Liberal Arts Faculty Teaching Award. He is also the recipient of the 2017 Dr. Michael Oliker Award for Outstanding Contribution to Philosophy of Education (awarded by the Society for the Philosophical Study of Education, USA). He has published six philosophy monographs and numerous peer-reviewed journal articles on such diverse topics as ancient Greek philosophy, phenomenology and hermeneutics, education, and film-philosophy. His most recent Routledge book, *Heidegger on Literature, Poetry, and Education after the "Turn": At the Limits of Metaphysics*, was co-authored with Elias Schwieler.

Part I

Eco-phenomenological issues

Transcending the *metaphysics of presence*

1 Introducing eco-phenomenology

Methods, problems, and proposed solutions

Prologue: the need for eco-phenomenology

We introduce what might serve as an epigraph for our study by turning to Rainer Maria Rilke, who in his *Letters to a Young Poet* offers the following profound – and, shall we say, "eco-phenomenological" – advice, urging him to "draw near to Nature. Then try, like some first human being, to say what you see and experience and love and lose."[1] The main and persistent challenge that eco-phenomenology faces is that humans on a global level *should* and, beyond, *need* – hence the sense of exigency – to demonstrate a new or renewed attitude toward the natural world (*Nature*) in order to arrest its looming destruction, to which, in many ways, we are heedlessly contributing. This is a "fact" grounded in scientific validation and corroboration as is evident in the most recent climate report released to the public on November 25, 2018, which is titled the *Fourth National Climate Assessment* and contains two volumes, with Volume I assessing the physical science underlying the study and Volume II focusing on the impacts and risks for the US. The conclusions are devastating and forecast a bleak future for not only the country, but also the world, the planet, if serious collective action is not taken immediately: "Global warming is transforming where and how we live and presents growing challenges to human health and quality of life, the economy, and the natural systems that support us."[2] It is the case, as the report authors warn, that in order to "avoid substantial damages to the U.S. economy, environment, and human health and well-being over the coming decades," humans must aggressively act *now* in an ecumenical, purposeful, and informed manner. However, although this report emerged from the Trump administration, the president himself stated flatly that he does not believe it, this because he dogmatically clings to an anti-scientific worldview and is a radical climate change/global warming denier. Here, although this lack of belief is frightening and flies in the face of logic and even common sense, it is the notion of "belief" that is of great interest to us in our pursuit of eco-phenomenological solutions to the types of issues contained in this important scientific document.

Let us recall Wittgenstein's brilliance as related to this issue of belief. When faced with a king who steadfastly anchors himself to the *belief* that the earth is flat, Wittgenstein asserts that all the scientists in the world presenting "facts" refuting the king's *belief* that the earth is flat would have absolutely no effect on altering the king's belief; facts would be powerless to change his mindset or, better, his worldview. Thus, what Wittgenstein ultimately shows is that in order to change the king's mind, his *beliefs*, as opposed to the presentation of facts, a total change to his entire worldview or overarching *belief-system* would be required – in terms related to our study, this would require a change to the king's *attunement*. Eco-phenomenologist David Wood states emphatically that the value of incorporating phenomenology into environmental philosophy lies in its potential to transform our mindsets, views, and moods, – i.e., *attunement* – working to reveal the "wealth of subtlety" contained in our *ordinary experience* of the natural world without "reductive schematization."[3] However, it is not the case that phenomenology only seeks to reveal the dangers of the schematization of original experience, in a purely *diagnostic* manner, it also, in a *prescriptive* manner, seeks to incorporate the insights gleaned from the understanding of the richness and complexity of our "lived" or pre-theoretical modes of *Being-in-the-world*, which phenomenology puts us back in touch with, in ways that *re-attune* us to new ways of living productively and ethically with the natural world. Because eco-phenomenology dives beneath the everyday modes of inhabiting the world with others – ways that are antecedent to the *schematization* of that experience – it "responds to the dullness with which we often *live* our ordinary experience."[4]

Deep Ecology and phenomenology

Deep Ecology

Deep Ecology was established by Arne Naess and highlights an expanded view of the human's spiritual consciousness through understanding its deep and ineluctable connection to the Earth. According to Naess, we are inextricably bound within the earth's web of life, and so it is inevitable that we should, in contradistinction to the current stance we have taken toward nature, "develop a broad and deep concern for life conditions in general."[5] According to Naess, since environmental problems can be directly traced to our anthropocentric behaviors, if we follow the principles of Deep Ecology, it is possible to transcend our anthropocentric attitude and move toward a "position of biocentric egalitarianism," and so potentially avoid "environmental catastrophe."[6] Deep Ecology is a form of environmental philosophy (*ecophilosophy*) focused on *normative* issues, expressing a "value priority system only in part based on results (or lack of results) of scientific research."[7] Naess claims that Deep Ecology should be separated off from ecological movements that exclusively embrace science and strict scientific methods, for it seeks to address more fundamental issues comprising the human-and-nature relationship, and so as opposed to "factual," it is speculative and provides "a kind of *sofia* wisdom," offering "value priority pronouncements *and* hypotheses concerning the state of affairs in our universe."[8]

However, Naess is clear that Deep Ecology is not philosophy in an academic sense, i.e., it is neither systematic nor ideological, but rather is both *descriptive* and *prescriptive* in nature and it shares these traits with eco-phenomenology.[9] The area of political philosophy is one of its concerns, for as Naess argues, the ability to reassess our current norms through the practice of Deep Ecology holds the potential to influence politics in and through a change of mindset on the part of those involved.

Deep Ecology rejects the human-in-the-environment model and adopts a *relational, total field model*, where all organisms are "knots in the biospherical net or field of intrinsic relations,"[10] i.e., the relationships are symbiotic and foundational, highlighted by *diversity*. Deep Ecology looks for ways to enhance the natural "potentialities of survival, the chances of new modes of life, the richness of forms," embracing the necessity of humans and nature to "coexist and cooperate in complex relationships, rather than [enhancing our] ability to kill, exploit, and suppress."[11] Symbiosis, as understood by Naess, inspires our identification with other forms of life, for they have an influence on our potential for our deepening sense of *selfhood*, i.e., "we increasingly see ourselves in other beings, and others see themselves in us," and in this way the "self is extended and deepened as a natural process of the realization of its potentialities in others."[12] Along with diversity, *complexity* is recognized, because it is argued that the attempt to understand nature in terms of a complex organic system will enlighten human beings about the complex biospherical relations and therefore inform the ethical ways we interact with nature, facilitating a deeper understanding of the effect we have on both enhancing and disturbing – and ultimately destroying – Nature. Deep Ecology embraces *biospherical egalitarianism*, which indicates that we should strive to acquire a deeper sense of respect and veneration for non-human life forms and the planet itself. This sense of veneration is linked to an *intrinsic* view of values, where nature is valuable independent of its instrumental usefulness for our purposes. For Naess, *biospherical egalitarianism* is bound up with the idea of "Self-realization" in a universal sense, where the "large comprehensive Self (with a capital 'S')" associated with nature – reminiscent of the Hegelian "Absolute" – "embraces all the life forms on the planet."[13]

If we examine the difference between Shallow Ecology and Deep Ecology on the issue of pollution we find that a "shallow" approach, according to Naess, seeks exclusively to purify the air while attempting to limit further pollution by proposing laws that regulate and restrict the use of technologies and modes of manufacturing that contribute to polluting the atmosphere. But in Shallow Ecology, such actions, guided by a latent and insidious anthropocentricism, are limited to confronting and battling pollution, with the central aim of perpetuating human thriving, health, and affluence in the developed countries. Importantly, in Shallow Ecology there is no critical thought regarding the ethical or philosophical ideas that underlie the vast amount of heedless and wanton pollution of the environment, whereas the philosophically inspired Deep Ecology is committed to "articulate the fundamental presuppositions underlying the dominant economic approach in terms of value priorities."[14] Deep Ecology focuses on the way pollution affects not merely the human denizens of the earth, but rather the entire biosystem, and calls for a "high priority fight against the economic conditions and technology responsible for producing [pollution]."[15] In short, in order to effectively combat the deep causes or motives responsible for modes of comportment that include embracing technologies causing pollution, we need to understand those motives by means of critical philosophical reflection. For example, as we discuss in Chapter 2, the project of *sustainable development* is traceable to a "shallow ecological" response because its *normative* concern for sustaining the health of the planet and the preservation of its vast resources has its origin in the preservation of *homo sapiens*, i.e., it is "anthropocentric in the sense that all its recommendations are justified exclusively in terms of their effects upon human health and basic well-being."[16] Sustainable development lacks the relevant philosophical perspective based on a set of norms grounded in the belief that nature is intrinsically valuable, which is consistent with a view toward *ecological egalitarianism*, a radically *eco-centric* position as opposed to an *anthropocentric* one.

Eco-phenomenology adopts a critical view of Deep Ecology, which as an *eco-centric* philosophy holds the potential to devolve into a form of *metaphysical* or *organic holism*, a view where the human being is not merely subsumed under, but is *sacrificed* to, the unquestioned dominant superiority of nature (*Nature* lives as *transcendental signifier*). Naess states quite correctly that the "earth does not belong to humans."[17] However, he graduates beyond this obvious and humbling truism by boldly declaring that when the human's "non-vital needs come in conflict with the vital needs of nonhumans, then humans should [in all circumstances] defer to the latter,"[18] and indeed, without falling victim to the *slippery slope fallacy*, many critics of Naess have arrived at the alarmingly disturbing conclusion that when such situations arise *humans* should be sacrificed for the superior good of nature's "Self-realization." As evidential fact, the ever-growing human population exercises an influence on nature because of the continued increase in the need for essential resources that are rapidly dwindling on a global level. Indeed, Naess states emphatically that the "flourishing of human life and cultures is compatible with a substantially smaller human population," and beyond this, goes on to add that the "flourishing of non-human life *requires* a smaller human population."[19] According to eco-phenomenologist David Wood, in no uncertain terms, this represents Deep Ecology's tendency toward *fascism*, in that Naess's claims could be marshaled in service of justifying "active human population reduction to accommodate the needs of other species, and that more broadly, the rights of individual humans are to be subordinated to those species."[20] It is also possible, along with Louise Economides, to read Deep Ecology as embracing and espousing an ethic wherein "individual organisms lack ethical importance, for they are merely temporary instantiations of enduring species," leading to a form of "holistic totalitarianism wherein the rights of individuals are automatically subordinated to the collective good."[21] As opposed to facilitating the potential for transcending "modernity's dualistic alienation of human beings from nature as a domain to be dominated in the name of civilization," it produces a "pathological form of identification,"[22] which is bound up with a pernicious form of holistic totalitarianism that forecloses the possibility of an ethics that is consistent with the *egalitarian* and *symbiotic* view of nature and the human being that Deep Ecology itself embraces and philosophizes.

Phenomenology

We begin with a general definition of phenomenology as a method or way of approaching philosophy that looks to first-person (subjective) experiences in order to describe such experiences in search of the *essential structures* that open the potential for meaning to emerge from that experience. However, this is not to indicate, as Dan Zahavi emphasizes, that phenomenology is merely about turning "the gaze inward to examine the happenings in a private interior sphere,"[23] and we explain this claim below. When speaking of *first-person* experience, we are referencing the phenomenological *intentional-correlation* between *human-and-world*, or in our case, human-and-*Nature*; the supreme importance of understanding this *correlation* will be addressed in Chapter 3 in our discussion of Graham Harman's *Object-Oriented Ontology*. David Detmer provides a basic understanding of phenomenology when asserting that phenomenology "seeks to describe the objects of experience and the acts of consciousness (for example thinking, perceiving, imagining, doubting, questioning, loving, hating, etc.) by and through which these objects are disclosed,"[24] and the scope of its inquiry includes all phenomena, e.g., aesthetic, ethical, and political. Phenomenology is most concerned with the world ("life-world") as it is given to us immediately in both our pre-reflective or pre-theoretical experience prior to the ways we conceptualize, categorize, or theorize about our experience. However, and this is crucial, phenomenology's project is not to seek an escape into the pre-reflective world of experience and remain there, nor is it concerned with simply providing detailed descriptions of this world. To do the former, according to Zahavi, would require no substantial methodological component, and "phenomenological studies would arguably be superfluous."[25] To achieve the latter would "simply reproduce the experience unaltered," which would amount to understanding phenomenology as a synonym for "phenomenality," and we find phenomenality in the literature of Proust, James, and Faulkner, but find phenomenology – as a philosophical approach and method – only in such thinkers as Husserl, Heidegger, Sartre, and Merleau-Ponty. Therefore, asking a question such as "What is pre-reflective experience like?" is *not* phenomenology's essential question, but serves only to open a vista into the far more complex *phenomenological project*. So, it is *not* that we are concerned with simply "seeking to attend to hitherto unreflected and unnoticed aspects and details of lived experience,"[26] but rather we are concerned with questioning how these experiences can potentially "provide us with a deeper understanding of fundamental ontological questions."[27]

This means that phenomenology works to "disclose, disentangle, explicate, and articulate [the] components and structures that are implicitly contained in the prereflective experience,"[28] and then to question the manner in which our theoretical attitudes giving structure to our scientific views "arise out of and change our pretheoretical Being-in-the-world."[29] In turn, this leads us to question how the pre-theoretical insight we glean from our descriptions might change the way we approach our theoretical projects, i.e., contributing to an altered relationship to science and technology. Importantly, this project is carried into eco-phenomenology with the goal to "understand the transition between our prereflective and pre-conceptual grasp of the world and our subsequent conceptualization and judgment about it,"[30] and this *connection* facilitates the rethinking and reassessment of our life-project in a re-attuned manner. Since phenomenologists vary regarding the use of either "pre-reflective" or "pre-reflexive" to name our pre-theoretical mode of dwelling, it is crucial to note at the outset that we choose in this book to employ the technical phenomenological term "pre-reflective," and view it as synonymous with "pre-reflexive," indicating two important things regarding the intentional mode achieved when the *natural attitude* is "bracketed": (1) It is a state that is not "reflective" in terms of formal, discursive thought (i.e., the theoretical view) and (2) The "I" or "subject" remains implicit in this pre-theoretical intentional state, in terms that are at odds with reflexive states, where the "I" is explicit in the thought process and at one with the direct object, as in our grammatical understanding of a *reflexive verb*. Most readers are familiar with the technical method of *epoche*-and-*eidetic reduction* that we encounter in Husserl, who, according to Zahavi, "is the only major figure among classical phenomenologists who consistently and explicitly made us of the epoche and the reduction,"[31] although it is the case that many contemporary phenomenologists – outside of the discipline of eco-phenomenology – embrace this Husserlian method, e.g., we encounter it in Vivian Sobchack's film-philosophy and J. Gordon Chamberlin's pioneering work in phenomenology of education. However, we introduce a more concise generalized version of such a method, common to many manifestations of eco-phenomenology, which might consist of *experience-description-reflection-analysis*, and which includes the search for patterns, repetitive motifs, consistencies, and connections present within the content matter of the descriptions that phenomenology seeks to wrest from concealment by making explicit the tacit (background) *meaning-structures* giving order to our Being-in-the-world.[32] These *meaning-structures* or *essential structures* (*essences*) of experience, in less ambiguous terminology, might simply be called "themes" common to the human condition, and we explore this notion below.

To concretize a common and recurring *ontological-existential* "theme" within phenomenology – be that in "existential" or in what is termed *Existenz* philosophy – let us briefly examine death, or more accurately, the phenomenological "theme" or overarching *ontological* structure of human *mortality*. Here, phenomenology seeks what is common or "essential" to mortality *qua* mortality, in terms of a potential ontological *meaning-structure*, rather than demonstrating a concern for death as a physiological or biological construct, categorization, designation, or fact, e.g., science with certainty can tell us that death is the cessation of functioning organs of sentient beings. Contrarily, through description and analysis, when *mortality* is approached as an ontological *theme* or *meaning-structure*, phenomenology demonstrates that our lives are stretched out temporally within the context of *finitude* with the potential for limited instances of human transcendence, and this ontological *meaning-structure* or *theme* gives form to aspects of the human condition that relate to and are bound up with other structures such as *spatiality*, *temporality*, and *communality*. It is within the *existential situation* of human *mortality* that meaning arises, for death is perhaps the most radical possibility of our Being; we cannot outstrip human *mortality*, but must rather live within it and struggle against its constant and looming presence. For example, death, in the phenomenology or *Existenz* philosophy of Karl Jaspers, represents one of several "ultimate situations" (*Grenzsituation*), or *meaning-structures*, which can neither be changed nor surmounted, only acknowledged and confronted.[33] However, this is not to indicate that all phenomenology is concerned with themes that can be labeled, with some caution, "existential" or even "ontological," for contemporary strands of phenomenology are now partnered with the philosophy of mind and cognitive psychology – although phenomenology is *not* a form of psychology itself – as in the phenomenology of Zahavi, Shaun Gallagher, and Daniel Dahlstrom, whose project is to understand human consciousness by studying "experiences as they are experienced in first-personal terms [in a way that] signals neither a fallback to a Cartesian paradigm of the mind nor the espousal of an antiscientific attitude."[34]

Phenomenology first emerges in Edmund Husserl as a rigorous and serious method for dealing with issues relating to consciousness and ways of cognizing and hence understanding the self, the world, and others. Although Husserl's philosophy transforms over the years, it is usually understood in terms of the following method and aims: Phenomenology is the search for the *eidetic* structures of the *things themselves* – this also includes the search for the a priori structures of consciousness as such (*transcendental ego*) or essential *subjectivity* – in which one engages in "bracketing" (*epoche*) experience, in terms of suspending everyday attitudes, dogmatic views, and scientific and psychological frameworks, and then one performs the *eidetic reduction* to reveal only the most essential structures (*eide*) of our experiences, which are grasped thematically in the absence of an empirical understanding of the world. Zahavi, who has published extensively on Husserl, describes the *epoche* and reduction, in the following succinct terms, as crucial elements of philosophical reflection that proposes to free us from

> our natural(istic) dogmatism and make us aware ... of the extent to which our own subjective accomplishments are at play when worldly objects appear in the way they do and with the vitality and meaning they have. By effectuating the epoche and ... reduction, Husserl ultimately came to embrace the view that reason, truth and being are essentially interlinked, and that the right place to locate objectivity is in ... the appearing world.[35]

Zahavi also informs us that in the milestone of twentieth-century philosophy, *Logical Investigations*, Husserl deals with such topics as the "defense of the irreducibility of logic" and the analysis of "part-whole relationships" and makes significant strides toward "showing that logical principles could not be reduced to psychological regularities." It was Husserl's focus on mental and intellectual acts as understood through a view of conscious *intentionality*, or consciousness as *object-directed*, that inspired Martin Heidegger to move in an "ontological" direction away from Husserl.[36] In *Being and Time*, Heidegger frames and adopts as a method what he terms the *fundamental ontology* of Dasein, which is focused on moods (*Stimmungen*) and attunement (*Befindlichkeit*), and so breaks with Husserl's method of *epoche* and *eidetic reduction*, which is dependent on the centrality of consciousness as a fundamental structure for understanding experience. In his early *magnum opus* (*Being and Time*) Heidegger sought the basic *ontologico-existential* structures organizing Dasein's mode of Being-in-the-world, bringing to our attention four basic ontological *meaning-structures* or "modes of Being-in": Falling, Discourse, Understanding, and Moods/Attunement. In Heidegger's later philosophy into and after the "Turn" (*Kehre*) of the 1930s, he abandons the practice of "philosophy" per se and adopts a far less academic mode of meditative "thinking." This represents the move from *fundamental ontology* and the exclusive focus on the experiences of Dasein as *transcendence* to consideration of other paradigms for the analysis of the *Being-event* or Being's presencing, which included for Heidegger an intensified study of textual sources, both philosophical and poetic.

Eco-phenomenologist John Llewelyn makes the interesting case that phenomenology essentially defies definition, if by "definition" we indicate an account of an essential nature, as in an explanation demonstrating propositional certainty. However, it can be understood through the idea of *family resemblance*, which includes "phenomenology of conception, phenomenology of perception, and the phenomenology that comes to exceed both conception and perception in the course of the history of [a] word's use."[37] This indicates that although various foci and forms or practices of phenomenology are not identical or reducible to a single issue or unified method, they share commonalities. For our purpose we examine these common traits or approaches and relate them to the crucial issue of phenomenology's relationship to eco-phenomenology:

1 Phenomenology brings our attention to the non-rational, non-theoretical (pre-reflective) dimensions of our existence, which include *affective modes* of understanding that hold as much importance and value for our lives as do formal *cognitive modes* of knowing;
2 Phenomenology opens us to an understanding of the world that is irreducible to the theoretical conceptualizations and categorizations we employ to organize and understand our "lived" experience of the world;
3 Phenomenology nurtures the view that the human, along with the world, is irreducible to any sets of facts that emerge from such disciplines as biology, anthropology, psychology, or sociology; and
4 Phenomenology, as related to existentialism, is concerned with the human's responsibility to its own autonomy, a trait identified by Mark Wrathall and

Hubert Dreyfus. For example, it is the case, in Sartrean terms, that the human is *condemned to be free*.[38]

Wrathall and Dreyfus also make the claim that phenomenology is primarily concerned with "providing a description of human existence and the human world that reveals *it as it is*, without the distortion of any scientific presuppositions."[39] Here, we must avoid misreading this claim; it is certainly not the case that phenomenology can provide sure and certain insight into the *things as they really are*, and so this statement requires further explanation and clarification.

As David Detmer points out, phenomenology, while appearing grand and ambitious in its aims and purposes, in another sense "is quite modest," in that while assisting us in seeing the *things we have already seen* more clearly through its dedication to "descriptive fidelity,"[40] it resists positing overarching and definitive theories that "would explain it all," or give us things as they are *in themselves*. What phenomenology can do, in a limited fashion as will be suggested, is to persuade us "to remove the confused, or speculative, or prejudicial assumptions to thought constructions that can serve to block our access to what is given in experience,"[41] as much as this is possible to accomplish through one or another form of "reduction" or *thematic analysis*. Here, we identify the limitations of phenomenology, an important concern for Graham Harman, who equates phenomenology with "continental philosophy," which he states is representative of "all present-day philosophy that takes its primary orientation from the phenomenological movement, whether in a positive or negative spirit."[42] Harman is rightly concerned with the danger inherent within any phenomenological claims granting *primary status* to the pre-theoretical mode of world-disclosure – which might somehow give us the things *as they really are in and of themselves*. As we see in Chapter 3 and our encounter with nature in relation to Harman's *Object-Oriented Ontology* and its unique philosophy of objects, to adopt such a phenomenological view as described above is to remain blind to what is indicative of *excessiveness* of the world and things of our experience, in the sense that the things of the world we encounter *transcend* or *outstrip* our best efforts to bring them to full disclosure, this despite the philosophical approach or method we adopt and practice. As Harman argues, and these terms and concepts are unpacked later for the reader, "in technical terms, the attempt to paraphrase objects," in the ill-fated pursuit of fully grasping their truth – and here we include the world of nature as an *object* as well as a system or collective of *objects* – leads to distortion, obfuscation, and translation, i.e., "always amounts to undermining, overmining, or duomining them,"[43] by doing *philosophical* and *interpretive injustice* to them. These notions will be thoroughly explored in Chapter 3, but here we briefly introduce these terms, turning to Harman who notes that as *forms of knowing*, an "object is undermined if we explain it in terms of [only] its smaller constituents, by way of a downward reduction"; we engage in *overmining* when treating objects as "needlessly deep or spooky hypotheses by comparison to their tangible properties"; however, *undermining* and *overmining* rarely occur in isolation, for they combine "in mutual reinforcement" in the practice of *duomining* objects, as found in certain strands of scientific materialism where we simultaneously *undermine* objects when treating "ultimate particles, fields, strings, or indeterminate 'matter' as the ultimate layer of the cosmos," but also *overmine* objects when adopting the view that "mathematics can exhaust the primary qualities of this genuine layer."[44]

Harman identifies two realms or levels of experience that concern phenomenology, expressed through "the difference between a rich, implicit layer of dark pretheoretical experience, and a luminous but impoverished plane of explicit awareness," which is the division between *theory* and *praxis*.[45] Harman argues that both *theory* and *praxis* "objectify" the world in unique ways, by *caricaturing* it or *reducing* it to a "small number of properties in comparison with its inscrutable implicit depth."[46] For example, imagine inhabiting a small and confined room with other people. If we are engaged in engrossing conversation, our direct attention is no longer on the room and its furnishings; we are not, therefore, theoretically objectifying the room. In this instance, as Harman points out, we should all be on "equal footing," in that we "ought to be simply using it in its pretheoretical reality."[47] But this is not the case, because even at the pre-theoretical level the room is already different for each person occupying it. For example, our physicality is unique and so we respond to the space of the room differently, for it is the case that our responses, the ways in which we inhabit the room, differ in terms of our attitudes and proprioceptive responses: Some might find it claustrophobic, others may determine it to be too cold or too hot, thus each of us is "already in a different mood, so even if we never objectify the room theoretically, each of us is already tacitly objectifying the room as 'promising' or 'gloomy.'"[48] This relates to the *excessiveness* and *inaccessibility* of all the things we experience, and to the complexity of the internal and external aspects of all things, which, as we will see, includes aspects of existence that are both *aesthetic* and *normative* in nature; and this reveals the multifarious and innumerable *perspectival* modes of the world's presencing for our interpretation and potential understanding. Phenomenology allows us to temporarily gain access to the rich and unplumbed depths of pre-reflective comportment so that we might glean and preserve some small aspect of that experience that has been revealed and clarified through our descriptive and interpretive activity, and then bring the experience back with us to enhance a growing and ever-more inclusive view of the world and our place in it. It must be noted that the exploration here of Harman's critique of phenomenology is not meant to suggest that the project itself should be abandoned or that its results are of no value or consequence; indeed, Harman claims that "phenomenology is one of the philosophies of human access par excellence – even more so than Kant's philosophy."[49] To reiterate, while this brief excursus should serve to reinforce the understanding of phenomenology's limitations, it is not intended to trivialize its potential benefits as we embark on our analysis of eco-phenomenology and nature.

The move to eco-phenomenology

Defining eco-phenomenology and introducing the scope of its concern

Following Brown and Toadvine, we adopt the label "eco-phenomenology" when referencing the environmental discipline that Zimmerman and Esbjorn-Hargens call *Environmental Phenomenology*. This unique form of environmental philosophy began in the mid-1980s when environmental thinkers turned to phenomenologists such as Husserl, Heidegger, Nietzsche, Levinas, and Merleau-Ponty. Eco-phenomenology "utilizes the methodologies of phenomenological inquiry to 're-see' ecological and natural phenomena freed from our habitual ways of conceiving and experiencing, thereby opening up new horizons of perception and action."[50] Eco-phenomenology is a unique form of eco-philosophy concerned with the current and continuing problems plaguing the planet that have been and continue to be – in great part – accelerated by humans wielding science and technology in the service of, to reference John Gray, the dangerous and hubristic belief in unbounded human progress at the expense of the environment, e.g., air and water pollution, soil and land pollution, climate change, global warming, deforestation, increased carbon footprint, genetic modification, destruction of marine life, public health issues, habitat destruction, ozone layer depletion, and natural resource depletion. These instances of destruction to the planet and its inhabitants, as Charles Brown and Ted Toadvine contend, should move us deeply, inducing an experience of "moral unease over what has been done, what we have done to nature."[51] It is here that we introduce the double claim guiding the practice of eco-phenomenology, as Brown and Toadvine articulate: (1) In order to better understand our current ecological status, the turn to phenomenology is necessary, and (2) "phenomenology, led by its own momentum, becomes a philosophical ecology, that is, a study of the interrelationship between organism and world in its metaphysical and axiological dimensions."[52] In light of Harman's warning regarding the limits of phenomenology, Monika Langer states that any phenomenological contribution will always and necessarily only be *partial*, and she also contributes unique thoughts to the issue of how phenomenology might enhance the field of eco-phenomenology, including the idea that, in addition to it representing a new addition to emerging integrated approaches to environmental philosophy, phenomenology is a discipline that also offers a necessary critical component, namely,

> phenomenological critiques of the traditional, dualistic ontology; alternative, non-dualistic phenomenological ontologies; detailed phenomenological descriptions to support environmentalists' claims concerning the primacy of lived experience; phenomenological descriptions showing the profound import of language – thus implicitly cautioning against the careless use of environmental discourse; and phenomenological insights that are particularly pertinent for environmental ethics.[53]

The turn to philosophy in environmental studies is thought to open new perspectives into these issues, and the inclusion of phenomenology, with its concern with ontological issues such as the study of ontological themes and human consciousness, can potentially open perspectives that differ greatly from abstract, theoretical views that highlight most specifically a *naturalistic* view, the goal being, in light of what we said about the phenomenological project, to transform our relationship to the conceptual and theoretical ways we disclose and interact with the world. As Brown and Toadvine observe, "One point of agreement among phenomenologists is their criticism and rejection of the tendency of scientific naturalism to forget its own roots in experience."[54] As we explore in Chapter 2, the modern era embraces *scientific naturalism* as the privileged mode of world-disclosure, and this is, according to Brown and Toadvine, an expression of our "myopic obsession with objectivity, our anthropocentric conceptions of value, and other legacies of Cartesian dualism."[55] It is phenomenology, and here we recall what was stated regarding the limitations of all forms of philosophical thought, that might suggest alternative modes of thought to broaden and deepen our view to the world, transcending the boundaries of our current limited perspectives, by opening a vista unto our potentially transformed and enlightened relationship with nature that is, at its core, a value-laden mode of earthly dwelling. The inclusion of phenomenology in environmental philosophy also holds the potential to foster interdisciplinary inquiries into the exploration of a renewed understanding of the human's relationship with nature, which in addition to ethics might also include aesthetics, henceforth putting in question the strict and limited boundaries that currently delineate and limit our understanding of and practices within various disciplinary modes of inquiry. This is a crucial theme explored within the *integral ecology* of Sean Esbjorn-Hargens and Michael Zimmerman. With the talk of *scientific naturalism*, which is not to be equated with adopting a "scientific attitude" and embracing empirical knowledge, it must be noted that eco-phenomenology neither disparages nor seeks to negate the valuable contributions of the sciences to the study of the environment, but rather seeks to integrate a non-objectivist view of nature and the human being into scientific study. Science is not the enemy of eco-phenomenology; rather eco-phenomenology attempts to foster a renewed attitude, or mode of *attunement* from out of which we comport ourselves, toward science and technology. As we explore in Chapter 2, it is radical *naturalism* or view to *scientism*, and *not* the trust in the reliability of scientific methods, that produces the problems eco-phenomenology seeks to address and potentially contribute to eradicating.

Iain Thomson offers an in-depth classification and critique of what he identifies as two forms of eco-phenomenology – *naturalistic* (eco-centric) and *transcendental* (humanistic-centric) – both of which are focused on the following duplicitous and interconnected issues in eco-phenomenology: All "environmentally-dangerous assumptions in question are 'ethical and metaphysical,'" and it is the case that the "offending assumptions most frequently singled out in eco-phenomenology are mind/world dualism and the fact/value divide."[56] It must be noted, however, that they also share a critique of traditional metaphysics and technology that harkens to, and in many ways is consistent with, Heidegger during the "Turn," as will be explored in Chapter 2. Let us briefly pause to examine a few of Thomson's conclusions. When eco-phenomenology turns to ether Husserl or Nietzsche, the ontological and ethical insights that are drawn from nature are equivalent to "facts," e.g., "Nietzscheans and Husserlians gravitate toward a *naturalistic* ethical realism," a form of *ethical perfectionism* "in which 'good' and 'bad' are ultimately *matters of fact*," they can be derived from an understanding of "good" and "bad" (ethical/unethical) in relation to those activities that enhance the planet – or the human's relationship with it – and those that degrade it. Conversely, eco-phenomenology focusing on Heidegger or Levinas espouses *transcendental* ethical realism, which is concerned with discovering "what really matters when we are appropriately open to the environment, but what we thereby discover are neither facts nor values but rather 'being as such,'"[57] which is a source of transcendental meaning that is neither reducible to or understood in terms of traditional values or facts. There are problems inherent in each these views, e.g., the former can devolve into a view that inherently privileges nature above the human, while the latter holds the potential to fall victim to the *fallacy of hypostatization*, which we will explore in forthcoming chapters, but here we note briefly that there are other forms of phenomenology that are incorporated into eco-phenomenology. Beyond this, in forthcoming chapters, we attempt to draw inspiration for our developing eco-phenomenological position from sources that represent neither eco-philosophy nor phenomenology, namely, from Graham Harman's *ontology of objects* and W. D. Ross's *ethical intuitionism* or prima facie *ethics*.

Nearly two decades ago, Brown and Toadvine observed the following: "Apart from a few hardened reactionary voices, the entire American mainstream from Disney to the White House now sings a green song."[58] However, at the time of this writing, things have changed dramatically, and we will lay out a sustained critique of the current organized move, as evidenced in current US governmental policies associated with the *Environmental Protection Agency* (EPA), away from a concern for the environment, in the radical direction of an attitude that could be summed up in the frightening adage – *In quest of corporate profits, the environment be damned!* Thus, in this day and age, perhaps now more than ever before, it is crucial that we turn to philosophical movements in the attempt to reverse this trend in destructive thought and behavior that leads to anti-environmental policies that heedlessly eschew regulations intended to limit various toxic pollutants from being pumped into the air and water by large profit-driven corporations and manufacturers. Since it is the case that the classic pioneers of phenomenology, as introduced above, do not espouse what might be termed a philosophy or phenomenology of the environment, it is the task of contemporary scholars to draw inspiration from primary phenomenological writings in order to infuse a nascent and emerging form of environmental philosophy with life. The reader will note that scholars from various disciplines have contributed to broadening and deepening the field of eco-phenomenology and continue to do so. For example, to provide but a short list, we encounter Christen Diehm writing on nature by incorporating the phenomenology of Emmanuel Levinas; Michael Zimmerman doing extensive work on Heidegger, Deep Ecology, and the environment; Lester Embree embracing Husserl in search of a *constitutive phenomenology* of the environment; Ted Toadvine writing on Maurice Merleau-Ponty's potential contribution to rethinking our relationship with the environment; and Monika Langer uniquely incorporating Nietzsche – in terms of Nietzsche as *proto-existentialist* – to reframe our relationship with the world and Earth. Indeed, there are also eco-phenomenological writings by philosophers breaking new ground in the field who do not align themselves with the "giants" of the continental tradition and instead adopt unique views of phenomenology, e.g., Wood, who philosophizes eco-phenomenology in terms of phenomenology's *rapprochement with naturalism*.

In line with what we have introduced, David Wood embraces eco-phenomenology as a practice that focuses on our interactions with and immersion in the world as *natural beings*, in ways that defy the findings of science as *naturalism*. Wood makes the following crucial observation that provides invaluable insight into the manner in which we are approaching eco-phenomenology: "If phenomenology is to be able to think about Nature, it must either rescue Nature itself from naturalism, or work out a new relationship to what it had perceived as the danger of naturalism. Or both."[59] In merging phenomenology and eco-philosophy a middle ground is sought, which indicates that phenomenology "opens up and develops an access to Nature and the natural that is both *independent* of the conceptuality of the natural sciences, and of traditional metaphysics."[60] Unpacking this claim is essential, for on first blush it appears that eco-phenomenology operates only at a level that stands at a remove and indeed seeks to radically break from science and technology. This cannot be so, for it is the case that the understanding gleaned from eco-phenomenological practice holds the potential to inspire a change to our mindset – or *attunement* – and, in a new and transformed manner, opens us to our intimate relationship with nature and the earth, grounding, guiding, and directing the renewed manner in which we approach or, better, ethically inhabit or dwell within the space of science and technology.

Certainly it is necessary to confront the negative effects of what Wood calls "traditional metaphysics," or what we identify as the *metaphysics of presence*, with both its Cartesian and Kantian origins. However, to reiterate, eco-phenomenology can never lose sight of the contribution it must seek to make to revealing and establishing a new relationship with science and technology. What we have termed radical *naturalism* or *scientism* is a problem engulfing the contemporary milieu, and we have observed that *scientism* is grounded in a modern *mindset* or *attunement* that is antecedent to our current interaction and engagement with advanced technology, i.e., it has already attuned the worldview into which we have been *thrown*, within which we have been acculturated and assimilated. It is the case that we hope to return enlightened from our eco-phenomenological or pre-reflective encounters in order to better understand how to carry and comport ourselves to nature, based on the pre-theoretical insight and *normative* understanding gleaned, which importantly includes our understanding and use of science and technology. To once again stress the following crucial point: we must avoid simply downplaying or, in the extreme, discounting the potential and power for positive change afforded by science and technology – the potential of making the earth inhabitable in a way that addresses the debt we owe it and responsibility we have toward it as a value-laden phenomenon in and of itself. For example, the use of alternative sources of energy undoubtedly requires science and technology; they are indispensable for this type of advanced environmental-friendly or *phenomenologically informed* environmental project. However, again, it is the attuned attitude toward the implementation of this technology in relation to nature that eco-phenomenology might transform for the better.

Eco-phenomenology also contributes to the recognition of perspectives beyond our limited personal view of things, revealing a trans-subjective/inter-subjective view of things, for in and through description, reflection, and critical analysis of our first-person perspectives a newfound understanding that points beyond the "subjective" emerges. For example, we only directly experience our small and intimate part of our involvement with nature. As Wood notes, if my tree is dying, I am the one that notices it, but there is always something happening that is beyond my perspective, and it is a fact that not only is *my tree dying*, but far beyond this, entire forests are dying and in continued danger of being destroyed; indeed the entire earth is in serious peril, and we might say without hyperbole that it is dying a slow, syncopated death, and we are, in the view of eco-phenomenology, in a way that graduates beyond our subjective experiences or perspectives, contributing to hastening its demise. Because it is beyond my subjective *glance*, this is not something I notice or am aware of; however, as Wood points out, this "glance is ripe for education,"[61] for the gap between what I see and what is concealed is something that I remain unaware of, but is something that eco-phenomenology might potentially bring to light, wrest from concealment. Staying with this theme, it is seemingly inconceivable that our personal behavior might somehow contribute to the irreparable damage being inflicted on the environment, for "to the extent that our consciousnesses of the significance of human action are resolutely individualistic, the collective impact of humans on the earth will fall beneath our radar screens."[62] In this instance, we encounter the gap between our individual actions and their contribution to the overarching collective impact we as humans have on the planet; here, we glide from our individual responsibility to our inter-subjective responsibilities as part of the ecumenical project to protect the earth from immanent destruction. Ultimately, eco-phenomenology awakens us to a transformed relationship between the human being and nature, a unique relationship that transcends the *metaphysics of presence*, which once revealed, informs the ethical attunement through which we transcend the gap between what is seen from a limited personal-subjective view and experience and what is concealed, namely, what is experienced collectively, manifesting a relationship that bridges the gap between the particular and the universal. From these foregoing remarks, we draw out five tenets expressing the aims and concerns of eco-phenomenology, as we understand it thus far, which guide and direct our inquiry:

1 There is a problem with the way we think and interact and hence treat/behave toward/respond to nature.
2 This problem is grounded in a tacit Cultural Framework (Conceptual Lens/Worldview) operative attuning humanity.
3 This Cultural Framework can potentially be transformed/transcended if we are made aware of its existence and its connection to the problem afflicting nature.
4 This calls for a reconceived, re-attuned understanding of nature and our relationship to it as it appears and is experienced outside of or antecedent to this Cultural Framework.
5 Once these "hidden" issues are wrested from concealment (via reflective phenomenology) and made explicit, the potential exists to change the ways we speak with, understand, and, ultimately, interact with nature.

Prolegomena to eco-phenomenology

Edmund Husserl, Martin Heidegger, essences, and the nature of understanding

When considering the potential for classic phenomenology to infuse eco-philosophy with new life, John Llewelyn reveals the challenges bound up with attempting to incorporate these phenomenological views in the pursuit of philosophizing nature through eco-phenomenology. Prior to exploring eco-phenomenology's goals and aims, we begin by considering the problematic issues it must attempt to overcome in its quest to open new avenues of thought in eco-philosophy. When merging phenomenology with eco-philosophy in quest of eco-phenomenological solutions, we must understand that in addition to "having an environment the human being is part of the environment of non-human beings."[63] Although we agree with Llewelyn's position, we prefer the incorporation of the terms "nature" and "earth" rather than "environment," simply because of the latent *naturalism* inherent in this formulation of the natural world, which although unintended, gives the disingenuous impression that the natural world is reducible to mere *organisms* functioning within an *environment*, and we have more to say about this issue in later chapters. The problematic issues we explore here are three in number: (1) The issue of the type and reliability of the knowledge or understanding that emerges from our phenomenological descriptions; (2) The issue of attempting to merge classical phenomenology as we find in both Husserl and Heidegger with eco-philosophy in the move toward eco-phenomenology; and (3) The issue of "essence," from the Latin *essentia*, in phenomenology and how this term might be understood as differing from the traditional metaphysical understanding of the reinterpreted

Greek term *ousia*, which holds the potential to lay the foundation for philosophical metaphysical *essentialism* – which is, of course, antithetical to the phenomenological view we embrace and espouse.

Phenomenology is not primarily concerned with understanding the issues it engages in terms of either rational or empirical "knowledge," for the concern is that these forms of knowledge are initially grounded more primordially: as Llewelyn recognizes, they arise "by abstraction from and dependent for [their] meaning on motivation by involvement in a concretely inhabited world."[64] Since it is the case that the claims and findings of eco-phenomenology neither take the *syllogistic* form nor are expressible through propositional locutions, as is consistent with the *Correspondence Model of Truth*, offering both sufficient and necessary reasons for their veracity, there is a movement away from the "positivist" model of the human and world in which the subject internalizes "knowledge" of that which is external in terms of *eidetic* re-presentation. Moving beyond Llewelyn for the moment, let us consider a form of knowledge that might be termed "phenomenological understanding," which emerges from our descriptions and analyses of lived experience, and which of course defies verification and validation that would be consistent with claims and arguments boasting necessary and sufficient conditions for their truth-value. We note here that Heidegger claimed that ontology becomes *as* possibility only by way of phenomenology, and phenomenology is impossible without hermeneutics. Thus, turning to Hans-Georg Gadamer, we encounter what might be termed "phenomenological understanding," which might be thought of as a form of not just *hermeneutic insight* but also *existential insight* – for we must be aware that Gadamer, much like Heidegger, deals with the issues of the *hermeneutics of facticity*, *human finitude*, and *ontology* – that emerges from the interpretive process as a form of affective and *normative* "wisdom" that the Greeks called *phronesis*, emerging originally from the Aristotelian distinction between *theoria* and *praxis*. Phenomenological understanding might be linked to our immersed experience of the world, for as Gadamer contends, it is primarily "directed towards the concrete situation," and so *phronesis* "must grasp the 'circumstances' in their infinite variety,"[65] and so this leads Gadamer to link *phronesis* intimately with *Erfahrung*, which is *experience* that facilitates understanding and is a perceptual and sensual encounter with phenomena as a form of *intuiting* as *aisthesis* ("knowing that is dependent upon sense perception").

To further explore these ideas in a manner that contributes to our view of eco-phenomenology's approach to *describing* the "things themselves" of our experience, in pursuit of a deepened form of *understanding*, we turn to Harman's description of "knowledge," "truth," and the essence of "philosophy." If we approach phenomena – and here we include humans, world, and organic and inorganic entities and things – with the pre-understanding or attuned mindset that they will always *withdraw* from our most ardent attempts to get to know them, they will therefore *withhold* their most intimate and *essential* secrets from us. It is often assumed that this leaves us with a thorny and unsolvable philosophical problem: If we cannot bring the knowledge of things to stand within propositional locutions – "clear statements of truth" – then we are left with no knowledge at all, we are left with, as in eco-phenomenology, meaningless and "vague poetic gesticulations," when in fact, as Harman contends, "most cognition takes neither of these two forms, as is clear from such domains as aesthetics, metaphor, design, the widely condemned discipline of rhetoric, and philosophy itself"; indeed Harman claims that philosophy, unlike science and mathematics, is "*not* a form of knowledge."[66] Harman looks to Plato's Socrates to show that "Meno's paradox" is solved by understanding that we are always on the *way to truth*, and that there are "cognitively valuable activities that are not primarily conceptual or discursive,"[67] much like we find both in Socrates' definitions – or, better, putative and malleable *descriptions* of the virtues he pursues – and in interpretation in the arts. Recall that, at least within the *early* "Socratic" dialogues of Plato, the so-called "knowledge" of the virtues does not "'capture' anything at the level of categories, concepts, predicates, or properties," but this is certainly not the equivalent of ignorance, for his persistent – but not *ironic* – claim to "non-knowledge" is *not just negative*, rather it is a form of understanding more akin to the experience of aesthetics, which reveals, through intimating, the things "obliquely, looking for paradox rather than literally accurate predicates as our entryway to a thing,"[68] in a way, we might say, that preserves the *mystery* inherent in all things, requiring that we not only accept but, beyond this, embrace the ontological *withdrawal* and *withholding* of the phenomena of our experience, of our inquiries. Indeed, our accepting and respecting the *withdrawal from knowledge* of phenomena is directly related to inspiring experiences that are *normative* in nature, as will be detailed in Chapter 4. The type of *understanding* we are seeking, Harman uniquely labels "unjustified true belief," although this is neither formal knowledge nor truth, if we understand truth defined in terms of providing "direct access to reality";[69] this will be explored in Chapter 3, focused on Harman's *Object-Oriented Ontology*, and we link this view of "knowing" as *unjustified true belief* with "intuition" in our ethical analysis of prima facie duties in Chapter 4.

First, turning to Husserl's phenomenology in its potential relation to eco-phenomenology, Llewelyn interprets the complexes of the *noesis-noema* and *cogito-cogitatum* as holding the key to answering the question of whether or not it is possible for Husserl's phenomenology to make a legitimate and valuable contribution to eco-philosophy. Llewelyn observes that in Husserl's phenomenology the *noesis-noema* complex of intentionality appears to retain the radical metaphysical divide between subjects and objects, and since "transcendental subjectivity constitutes objectivity," it is possible to state that if objectivity is a way of being afforded to the subject, then objects (*Gegenmstande*) of the natural world become secondary, they become "subjected to the subject, perhaps subjugated by it."[70] Natural objects are then understood and valued only according to the "subject's environment," and even if the natural world of objects is not radically "enslaved to the human ego, the human ego seems to be accorded a certain priority," and if we push this line of thought to the extreme, we encounter the exact type of *anthropocentricism* that forecloses any talk of a view to nature that would be consistent with Deep Ecology or, in relation to our concerns, eco-phenomenology. Thus, this interpretation of Husserl's phenomenology would contribute only to a *shallow* form of ecology as discussed earlier. In addition – although Llewelyn does not consider this issue – even if we properly treat the *noesis-noema* complex in terms of: (1) The operations of the conscious mind (*noesis*) as it is discerning meaning and (2) The meaning that is given through the act of perceiving the objects (*noema*), while keeping in mind Husserl's insistence that the *noema* is not a real object *simpliciter* – physical object – but an ideal entity inspired by but set apart from the physical object, it is still possible to draw out a metaphysical and *metaphysical/dualistic* picture in which this form of conscious intentionality remains trapped in an interpretation privileging the subject's internalization of the world in knowledge occurring through acts of *eidetic* re-presentation.

Second, considering Heidegger's phenomenology and the notion of *in-der-Welt-sein* (Being-in-the-world), the case can be made that early Heidegger also falls victim to the charge of latent subjectivity in his phenomenology, i.e., that the *fundamental ontology* of *Being and Time* privileges the human subject, and this is indeed a legitimate critique Heidegger himself raises in "Letter on Humanism," which makes it appear that Heidegger is embracing the technical phenomenological understanding that Dasein's *transcendence* and *temporal horizon* serve as the *condition of possibility* for understanding Being. The crucial implication of this reading, as Llewelyn contends, is that it determines the manner in which Dasein – whose Being is always a primary issue – inhabits and indeed creates its world in terms of instantiating a *referential totality* of meaningful activities, which is reducible to "a totality of utilities and disutilities, goods or means to the discovery or production of goods that are expected to answer human needs or desires."[71] In this way, just as in Husserl, early Heidegger's *fundamental ontology* rises only to the level of a *shallow ecology*, which might be related to *sustainable development*, in that although we might work to preserve natural resources necessary for facilitating human survival, and this might include a concern for non-human entities, this concern would be dictated by needs and desires of the human being. Llewelyn also talks briefly about a quasi-religious reading of Heidegger where Being is equated with God or *godhead*, and here we encounter a problem related to and emerging from Heidegger's later philosophy or *thought* of the "Turn" – the move from the subject-centered analysis of Dasein to an analysis of the truth-of-Being and Dasein's relation to it – which might be understood as the *fallacy of hypostasizing* Being. As Richard Polt also observes, this amounts to reifying Being as if it were an ontological object – a *transcendental signifier* – "turning it into some hyper-entity or divinity that is calling us, manipulating us, or commanding us."[72] Despite later Heidegger's popularity in both Deep Ecology circles and eco-phenomenology, this form of ecological philosophy holds the potential to devolve into a totalitarian or fascist view of nature which, as Llewelyn rightly notes, "sacrifices beings at the altar of Being";[73] or we might say, in relation to eco-philosophy, that Heidegger's later philosophy opens the potential to *sacrifice humans at the altar of nature*.

We now pause to reflect on a crucial issue in the practice of phenomenology that deals with the use of the term "essence" (*ousia*), and it is not just Husserl who talks of essences, for Merleau-Ponty observes that "phenomenology is also a philosophy that puts essences back into existence [but] does not expect to arrive at an understanding of man and the world from any starting point other than that of their 'facticity'."[74] Essence and its use, as phenomenologist Max van Manen points out, emerge from the distinction between a thing's *whatness* (*quiditty*), the essence of a thing, and its *thatness*, its mode of existing, which translates into the distinction "between *Grundwessen* (basic fundamental essence) and *empirisches Wesen* (empirical essence),"[75] e.g., there is an empirical essence of particular philosophers and a fundamental or ideal essence of the Philosopher as the *eidetic* structure that living (empirical) philosophers instantiate. The problem inherent in such an understanding is the danger of slipping into a form of *idealism* or *metaphysical essentialism* – which harbors *epistemological essentialism* – where essence might be understood in terms resembling the transcendental, immutable Forms, as in a doctrinal-idealist reading of Plato which, as related to the core of phenomenological practice, diminishes the primary importance of lived or perceptual experience, degrading the human's visceral terrestrial existence – which is pejoratively labeled "appearance" (*mē on*) and deemed inferior – in favor of a superior transcendental and transcendent "reality" – referenced as Being (*on*). In addition, the term "essence" can be traced back to the Latin mistranslation of the Greek word *ousia* that Heidegger brings to our attention, and this noun is based on the feminine present participle of the word *einai*, meaning, "to be." It is related to our English "being," but it has been mistranslated and hence misunderstood due to the Latin rendering of the word as *substantia* as related to *essentia*.[76] Thus, we accept its connotative and denotative force as referring to "substance," which gives the illusion of something tangible, an entity that is hypostatized as "thing-like." To avoid such misunderstandings, Llewelyn stresses that if phenomenology, and by extension eco-phenomenology, still seeks to retain the term, "essences will have to be more flexible," which indicates that the "essences that phenomenology is said to describe could still be more open-textured."[77]

In pursuing this line of thought regarding essences introduced but not pursued by Llewelyn, we look to the phenomenological research of van Manen (*Researching Lived Experience*), which is "determined by the question of the essential nature of lived experience: a certain way of being in the world," a concern that is twofold in nature – there is a "preoccupation with both the concreteness (the ontic) as well as the essential nature (ontological) of lived experience."[78] It is possible to think of the "essential nature" of lived experience as it is inseparable from *meaning*, in terms of ontological "essences" representing the Sense (*Sinn*)-giving-making structures that provide form to our everyday "lived world" experiences, and we instantiate or *live-out* these structures when comporting ourselves in *praxis*. However, in line with our phenomenological analysis, in our everyday modes of comportment these "meaning-structures" are concealed from our thematic awareness. Van Manen certainly has this in mind when pursuing "structures of meanings," which he calls *ontological themes* that provide structure and form to the lived-world, and it is this "grounding level" of human existence and this realm of pre-reflective experience that phenomenology studies in order to reveal its "thematic structure."[79] Van Manen considers a variety of topics or "themes" emerging from and directing the practice of phenomenological-ontology, e.g., "fundamental themes such as 'life,' 'death,' 'otherness,' 'meaning,' and 'mystery,'"[80] for these fundamental ontological themes make experience of the human condition possible in the first instance. However, van Manen selects four main themes, or structures, as guides for reflection in phenomenological research – namely, *lived space* (spatiality), *lived body* (corporality), *lived time* (temporality), and *lived human relation* (*relationality* or *communality*) – and these ontological structures/themes are directly relatable to the human's interaction with nature, as we later explore in Chapter 5.

Let us explore in a bit more detail the concept of theme(s) in van Manen by exploring the analogous relationship between themes and the idea of "essences" along with considering the type of language, or form of *naming*, that communicates themes in our phenomenological research. According to van Manen, themes serve a dual function: (1) They "give control and order to our research and writing," and more importantly for our purposes, (2) they "may be understood as the *structures of experience*," and so when we "analyze a phenomenon [experience] we are trying to determine what the themes are, the experiential structures that make up that experience."[81] As stated, themes such as *space, lived time, lived body*, etc., are not hypostatized Forms (*eidai*) or objects; they are in fact structures of meaning or sense (*Sinn*) that we ultimately reveal in and through our descriptions and "name" in the sense of the German *nemen*, where we name or mention something but do not represent it in a way that makes it fully present. Naming is more of a gesturing, suggesting, and intimating than the establishment – through a naming that attaches a categorizing label – of an essential thing or category to which we attach rigid and indelible characteristics. As Michael Steinmann observes:

> A "name" does not say that something *is* this or that. What it says is that *it can be called this way*. Names are meant to evoke something, to call it into the fore, by still preserving a sense for its independence and the untouchable ground out of which it emerges.[82]

So, in relation to the problem we have identified, when van Manen speaks of capturing the *eidos* or essence of something through thematic reflection, he indicates that *eidos* is a theme, or as we have indicated, a potential Sense (*Sinn*)-giving-making structure, which requires phenomenology to reveal and the hermeneutic to interpret. Themes are born of our primordial desire to make sense of experience in order to give shape to the shapeless, to arrive at an understanding of the things we are inquiring into, albeit in a way that is always incomplete and in need of further inquiry. According to van Manen, addressing themes is bound up with "*invention* (my interpretive product), *discovery* (the interpretive product of my dialogue with the text of life), *disclosure of meaning* (the interpretive product 'given' to me by the text of life itself.)"[83] Thus, when we talk of capturing the essence or *eidos* of phenomena through the practice of phenomenology, this indicates that we are dealing with "themes" as Sense (*Sinn*)-giving-making structures, which require, to reiterate, phenomenological *description* to tease out and the hermeneutic to interpret within our analyses. Contributing to this reevaluation and reinterpretation of "essence" in phenomenology in terms that enhance van Manen's view, Lingis, when writing on the *transcendent essence of things*, recognizes that it is not from the history of phenomenology that we inherit the misuse of "essence," rather this is traceable to classical epistemology, which conceives of "essence as ideality, something self-same that transcends, or recurs in the same in, any moment and any place," which indicates in an illogical and contradictory manner that for something to be the same to all of us it requires what does not exist, namely, an "impersonal or interpersonal transcendental ego."[84] Rather, in terms that invoke a powerful sense of musicality and the literary, harmonizing with van Manen's notion of *ontological themes*, Lingis rightly emphasizes that the

> sensuous essence for a perceived thing recurs in all its visual, tangible, resonant, and gustatory aspects, but not in the way that one same ingredient occurs in different compounds nor in the way that a law or principle recurs.... It [the essence] recurs in the way that a musical theme is one in all its variations, or in the way that one style is found in all the novels of an author who speaks with his or her own voice.[85]

Insuper Husserl and Heidegger

Maurice Merleau-Ponty, Alphonso Lingis, and interrogative ontology

We begin with an elucidation of "method" in eco-phenomenology prior to our analysis of Maurice Merleau-Ponty and Alphonso Lingis, for as Brown and Toadvine observe, although there have been "methodological divergences" over the course of phenomenology's first century, phenomenologists have continued to share the rallying cry first introduced by Husserl himself: *To the things themselves!* For the most part, eco-Phenomenology does not model or employ the technical methods of phenomenology as practiced by either Husserl (*epoche/eidetic reduction*) or Heidegger (*fundamental ontology*). Indeed, there is no universal agreement on the method adopted in the practice of eco-phenomenology, but there is a "general tendency to start from experience – here taking this term in a broad sense," and this already demonstrates a basic "convergence of phenomenological method with the concerns of contemporary environmental thought."[86] Let us consider a basic method of phenomenological *description* of the life-world that is first made possible when we hold in abeyance the technical structures consistent with scientific methods and systematic modes of theorizing.

This, we might term "bracketing," in a non-systematic-technical sense, a practice by which we "loosen" the ties with the world, through a "suspension" that instantiates, in the words of Erik Matthews, the "position of the 'perpetual beginner,' like a child who has learned nothing, can take nothing for granted, and so confronts the world of experience directly and naively, in wonder."[87] To achieve this state, in the words of Brown and Toadvine, is to reach a living mode of "ontological" or "metaphysical neutrality," which situates us "in a position to see that the world and the things in it are only given to us through the interpretative and meaning-bestowing (meaning discovering) function of our intentional acts."[88] This conscious or *immersed-intentional* state, as Wood points out, which marks a recovery of our *originary* immersion in and engagement with the *things themselves*, is not a return to a mode of "pure presence," rather it is a return to a world of experience that is active and "deeply complex and stratified," which opens up our lived immersion in the world of our engagement with other humans, non-humans, and as is our concern, the *things of nature*. This experience opened and afforded by eco-phenomenology, according to Wood, inspires an experience of the world that is essentially "normative" or value-laden. Lingis, whose important work on the *normative* imperatives in things, objects – and, as we argue, nature – will be explored, claims that beneath the lay of things as organized by the *naturalistic framework*, there is a lost and forgotten realm of human existence, highlighted by our "immersion in the elemental luminosity, sonority, a clearing upon the supporting earth, atmosphere, warmth and night," where things and objects take form as "salient and sensible, which only then mesh with the pathways of our practical worlds," and are present to and emerge from the interaction and commingling between us and the sensuous *things themselves*. Uniquely, for Lingis, phenomenology puts us in contact with, albeit obliquely, the "ends ordering the environing fields [that] function as directives, indeed as perceptual imperatives,"[89] and these imperatives, as we discover in Chapter 4, are *normative*; they are ethical in nature.

We have already discussed the ineluctable relationship between phenomenology and ontology, and now we extend this notion by invoking the phenomenology of Lingis, which draws inspiration from both Merleau-Ponty and Immanuel Levinas, to move in a unique direction. In Lingis we encounter the interconnectedness of ontology and the *interrogative mode of questioning-knowing*, which proceeds in the manner of *ontological interrogation*, which is a "form of thought that discovers that *the existing world itself exists in the interrogative mode*."[90] This type of questioning is not expressive of human ignorance or the privation of knowledge that could be overcome when answers are acquired, which then forecloses the path of further inquiry, or renewed and continued questioning. Rather, as Lingis asserts, the phenomenological question is a mode of Being-in-the-world that instantiates a "relationship with a world that forms and remains a horizon, a field of contingencies, potentialities, perspectival deformations," which at once discloses, conceals, forms, deforms, and remains "staggered out in depth."[91] Questioning is a process that is understood as unfolding in the mode of the *interrogative*, in relation to the *infinitival* mode, and we might say that it retains contact with the things themselves, and in a syncopated response to the very nature of their essential unfolding, "remains at a distance, contracts and dilates its presence, contests and negates its own course."[92] The ontological understanding of the things phenomenology seeks is irreducible to a relationship between subject and object, i.e., one of "adequation" or correspondence through what amounts to knowledge assuming the form of *eidetic* re-presentation, which wrongly indicates that asking the so-called "well-formulated" questions will eventually produce the justifiable and valid manifestation of "truth." The questions that are consistent with Lingis's view of *interrogative ontology* do not have definitive answers and, as stated, cannot authentically be closed down, and are indeed difficult to formulate due to the malleability, the swell and recession of what phenomenology attempts to momentarily bring to stand in a unique interrogative form that harbors elements of the ontological. However, as Lingis contends, perception in its essence assumes the form of questioning that manifests the "quest after things that are there from the start – from the beginning there is contact; they are where they are, but are also radically where we are,"[93] that live and arise from out of the vicissitudes and vacillations of our perceptions that are "marked out in the continuous fabric of being, of 'flesh,' of which we too are part."[94] As Lingis claims, original or *originary questions* cannot ground a "discourse in the grammatical form of interrogative sentences," they cannot give rise to truths that are expressible in "formulas, where unknown terms are determined by relationship with known terms."[95] However, let us consider some examples of the kinds of questions Lingis's phenomenology might inspire eco-phenomenology to consider when searching for a renewed and reinvigorated relation to nature. Below, we consider what might be termed a *primordial* and *attuned* relationship to a river, which we shall envision as afforded by the entryway into the pre-reflective realm and the subsequent experience of what we later elucidate as an *attuned* "river-experience" or *river effect*:

> How does the river appear to us in its brute and naked originality when we have suspended the conceptual lens of naturalism? How do we experience the river in what might be understood in terms of a primordial experience? How do we interact with it when simply standing in its presence? Is there an experience of involvement that lies beyond the aesthetic effect of the river that might move us in the face of its beauty and majesty? Or, might an aesthetic view capture something that remains elusive and is absent from other ways of viewing? How does the river interact with its surroundings? Is the river's essence reducible to a hydro-power source, or is the river's Being, in its most original essence, something beyond this? What might it want us to know about it? Is it a thoroughfare or obstacle? Does it appear as value-neutral or value-laden? Does it exist to serve our needs, wants, and goals, or for some other, perhaps unknown, and hence unnamed, purpose? Is it dependent upon us or are we dependent on it? Or, are we, in some important manner, in some degree or form, co-dependent?

Merleau-Ponty and the immersion of the human being in the world

Merleau-Ponty rejects the understanding that every experience is reducible to either a conscious act or the object of a conscious act. Vincent Descombes links this view to Jean Paul Sartre's understanding of the division between the *in-itself* (thing/material entity) and the *for-itself* (human/conscious intentionality), and it is the middle space or "terrain of the '*entre-deux*'"[96] that Merleau-Ponty's phenomenology inhabits. This is because of his notion of *embodiment*, a form of Being-in-the-world – or total immersion in the world – that is irreducible to either pure consciousness or objectiveness, for all consciousness is bodily in its essence and, further, is inseparable from the world or nature – here, we might invoke Nietzsche's claim that it is in fact the *body that thinks* and *not* the mind. For according to Merleau-Ponty, and this notion is what is called "Flesh (of the World)" – *la chair du monde* – in his later work,

> my body is made of the same flesh as the world (it is a perceived), and moreover that this flesh of my body is shared by the world, the world *reflects*, encroaches upon it and it encroaches upon the world.[97]

There is a holistic inter-relationship between the human and the world, or nature, and this relationship is one of "transgression or of overlapping,"[98] unfolding through an intimate dialogue between the human and nature, where corporeal intentionality merges with, because it responds to, the solicitation of the world, or what we shall come to understand as the "call" or "address" of *Nature*. Thus, against what Husserl identified as "transcendental ego," the quest to thematically arrive at the structure of consciousness, which facilitates our interpretation of meaning through our experiences, Merleau-Ponty's understanding of phenomenology is not concerned with revealing timeless a priori structures of existence as would be consistent with philosophy emerging from the Western tradition of metaphysics. Thus, the critiques of latent *subjectivity* and *idealism* that we addressed earlier in our discussion of Husserl and Heidegger do not apply to the phenomenology of Merleau-Ponty, and so the potential exists, when turning to Merleau-Ponty, for eco-phenomenology to break open a space for a unique phenomenological view of the intimate relation between the human and the world (of nature) that outstrips the charge of *anthropocentricism*. When eco-phenomenology embraces Merleau-Ponty, as in the work of both Toadvine and Langer, it confronts and assuages the tendency to "reduce and to reify the dialectic of being-in-the-world into a meaningless collection of inert objects observed by an unsituated spectator-manipulator."[99]

In relation to what was revealed by Llewelyn, it is Merleau-Ponty's drawing from Husserl's later thought, and attempting to move beyond it, that contributes to a new conception of phenomenology that is related to the "embodied" human being as immersed in the world as a "lived world." This *pre-given* or pre-reflective world is always already present prior to our experience and it is a phenomenon that defies our attempts to "bracket" it completely in the phenomenological reduction. What we can do, however, is bracket out theoretical *explanations* of our lived experience, which places us in the position to consider *descriptions* of this experience, and as Eric Matthews observes, here the aim of phenomenology is "not the achievement of rational insights into the 'essences' or necessary structures of experience, but a deeper understanding of the meaning of our theoretical activities through grasping their roots in ordinary lived experience," namely, "it moves from the realm of 'rigorous science' to that of 'radical reflection,'"[100] and as we have seen, Zahavi described this process as holding our *dogmatic* views in abeyance. We must note that when Matthews references "necessary structures" and "essences," he is not invoking the traditional metaphysical idea of categorical transcendental *eidetic* structures, but rather referencing the basic thematic "meaning-structures" that we introduced earlier in relation to van Manen's phenomenology. Thus, the *radical reflection* described by Matthews is indeed focused on teasing out *themes* (meaning-structures) that emerge from our *descriptions* of lived experience. And further to this point regarding our talk of "essences," as Matthews emphasizes, when Merleau-Ponty references *putting essence back into existence* he means that the human and its world should be understood in terms of the intimate relationship they share, and that to understand them requires "general concepts [*themes*], but these concepts and their analysis are not the central concern; rather, it is the relation to the world that gives meaning to these concepts [Sense (*Sinn*) meaning-giving-making structures] that is the primary interest of phenomenology."[101] Indeed, in terms that are nearly identical, Matthews claims that phenomenological understanding, in the search for the exploration of meaning, consists in "finding the *intelligible pattern* in some variety of human behavior or our response to the world,"[102] and this process, as Matthews asserts, leads to a deeper awareness of our intimate involvement with the world (nature) and reminds us of

> the impossibility of marking out a privileged domain of "inner experience," for by loosening our normal ties with the world and so coming to *wonder* at our relation to the world, we become more conscious both of the world's independence of ourselves, its simply "being there", and of the dependence of our own existence as conscious beings on our situation within the world.[103]

There can be no consciousness of a world without a world in the first instance; there is no inner sanctum of subjective thought outside of or separated off from an existing world of objects and entities giving rise to our thought, providing it with *content*. For here, we encounter the very antithesis of both *solipsism* and the latent *subjectivism* we identified in Husserl and Heidegger. In line with this thought, it is also the case that we do not encounter a cold, lifeless, value-free objective reality, but rather we encounter things in ways where the meanings inherent within them inspire a sense of meaning within us, within our involved dealings with them. Thus, it is possible to state that meaning and systems of meaning emerge through interpretive activity on the part of the human, but it is also the case that *meaning* and *systems of meaning* already exist, are already operative in the world itself. What becomes our *world* is the way in which the phenomena of our perceptions get organized in relation to our specific projects, but this does not indicate that we fully impose our meaning on the world – in terms related to an *anthropocentric* view – rather,

> the world comes into being as "meaningful", and we come into being as "subjects" standing in the meaningful relations to it, in the same interchange between ourselves as being in the world and the world in which we have our being.[104]

This importantly indicates in that we neither exercise total control over our perceptions, nor do we have complete control over the world, over the way in which the world shows up for our appropriation. Rather, according to Merleau-Ponty, nature appears in such a way that it is both within, and so available to, my intentions, but it is also there despite my autonomous intentions, e.g., when standing at the foot of a mountain range, "whether or not I have decided to climb them, these mountains appear high to me, because they exceed my body's power to take them in its stride."[105] Merleau-Ponty claims that because "[we] have hands, feet, a body, [we] sustain around [us] intentions which are not dependent on [our] decisions and which affect [our] surroundings in a way which [we] do not choose."[106] In light of the situation between human and world just sketched, all phenomenological reflection "must be provisional," and as related to what was introduced above in relation to *limitations* and the elusiveness and *excessiveness* of things, Matthews observes that "human beings are faced with a reality that is, in Merleau-Ponty's word, 'inexhaustible.'"[107] Although we make our world human in and through wresting various meanings from concealment, "paradoxically, an essential part of its being a human world is that is stretches indefinitely beyond any human attempt to encompass it."[108]

The radicalized phenomenology of Alphonso Lingis

The descriptive method and concern for the "things" of the world

We conclude this chapter by focusing on the phenomenology of Lingis, a well-known, respected translator of Merleau-Ponty and Immanuel Levinas, who philosophizes embodied intentionality as *corporeal intentionality* – *Empfindnisse*, hyletic materiality, kinesthetic sensations – which is a technical notion in his early work that is carried forward and repeatedly intimated throughout his corpus, but no longer formalized as a method in his later writings. Lingis graduates beyond the latent subjectivity present to both Husserl and Heidegger as consistent with the eco-phenomenological project concerned with transcending anthropocentrism. In Lingis, we are introduced to a unique view of *intentionality* in which every intention is directed toward objects or *things of the world* in terms of a movement or posturing that instantiates an *attitudinal orientation* toward the things we encounter. Thus, inherent within the *correlationism* we find in Lingis there is an intimate emotional component in that we are moved by and through our interaction with the objects of our concern, but interestingly and importantly, the *directionality of meaning* is reversed from that of philosophy's traditional understanding of *correlationism*, which is grounded in a decidedly anthropocentric view of the world that places the human at the center of the universe (from Human⇒world⇒*Meaning* to World⇒human⇒*meaning*). According to Lingis, our sensibility is dimensional and extends across time and space, and it receives "the sensuous within itself by actively taking up its vibrancy and internalized microstructure"[109] as we actively conform to the shifting and ever-changing contours of the world of which we are a part, within which we are immersed. Thus, when we recognize and understand the appearance of things of the world, this is not knowledge of "an ideal unity of aspects which would be reached by an intentional act transcending the sensible toward the ideal," but rather "the convergence of all receptive powers of the body upon it," which is "effected by the synthetic, or synergic, unity of the body – schema."[110] This indicates that the underlying structure of conscious intentionality is grounded in an anticipatory receptivity and openness that allows us to "conform to the style and grain of the things and thus receives within itself the patterning of their structures."[111] Importantly, this notion cannot be understood in terms of an external object presented to or re-presented within an internal subject, or "consciousness facing a *noema*," but is instead an "occurrence within a vortex," which is the "specializing-temporalizing" of the flesh,[112] for the sentient is "of the same sort of the being it knows, sensible like it,"[113] and so it is suggested that intentionality is the "receptive movement of a sentient being whose metaphysical constitution is like that of the being perceived – sensible like the world."[114]

Here, we encounter a radical understanding of perception that harkens in analogy to the pre-Socratic mystics, where in relation to "knowing" and "understanding" a world – a kinesthetic phenomenon that is also *synesthetic* in nature, rooted in the visual, auditory, tactile, gustatory, and olfactory – it is possible to invoke, for the purpose of elucidation, the phenomenon of *homoiosis*, or as we find in the ancient *physikoi*-poet Empedocles,[115] a form of *internal-external* assimilation or *physical-physiological* communion wherein *like is known by like*, which as related to our themes indicates two crucial things: "The knower cannot know an object without some type of identity of elements between them, and in knowing something we also, at the same time, become more like it."[116] This phenomenon is philosophized by Lingis, who asserts that we "know" the world through a perceptual process that unfolds in the duplicitous movement where we assimilate to the world as it simultaneously assimilates itself to us; but we must be clear that for Lingis, assimilation is *not* full "appropriation," rather it is an "apprehension" of the *things of the world* and the "circumscription of their forms," and in terms of opening the possibility of understanding, through the interpretive hermeneutic, our Being-in-the-world, the emergence of subjecthood, is essentially an "exploration of our forces" into the forms emerging from our conscious and interpretive interaction with the world.[117] This process reveals that the inner essence or *real formula* of a thing is never fully grasped, acquired, or exhausted, but rather serves as a continued "task for an exploration,"[118] and that the process of perception

> is ordered by the ordinances things realize, and we as perceivers realize what we are through the styles of postural integration they induce in us and in the images they project back on us of the way we look, hear, and feel to them.... It is the clay of our own body, dust that shall return to dust, that knows the earth and knows itself as terrestrial. It is the liquid crystals of our eyes that are turned to the stars as to eyes of the night.[119]

For example, Lingis describes the manner in which we come to know our bodies in the encounter through perception with the other's body, and this includes objects or non-human entities that populate the natural world, in the following terms: First, we have neither absolute nor conceptual knowledge of the *essences* of real things, and second, we do not possess or acquire an understanding of the technical and physiological processes that "regulate the position and focus of our limbs and organs."[120] Rather, we have an intuition of *how* to position ourselves before the things we encounter in what amounts to a *sympathetic response*, for in "perceiving the outer forms of the others, we capture in our postural schema the corresponding inner lines of their postures and movements," in addition to their "tensions and orientations,"[121] and to do so is to internalize their motor diagrams and hence to "quasi-perceive the visible, tangible, and audible form of ourselves turned toward them."[122] To reiterate a point introduced above, our immersion in the world and nature presupposes that our intentional relation to the *things of the world* is emotional and intuitive as opposed to rational or theoretical in nature, and this relationship, as we develop in Chapter 5, is such that we are open to and conform ourselves with the *things of the world* when listening and responding to their call or address, speaking in a distinct (reticent) manner unique to them, which demands an appropriate and ethically fitting rejoinder from us. For example, as Lingis observes, as we sing, on a hike through the woods, and "shout to the mountain cliffs we hear how we sound to the murmuring trees and in the great ear of the mountain."[123]

Ultimately, the turn to Lingis's phenomenology, in its potential relation to eco-phenomenology, will take the "form of reflection on the primal unity of the '*natura naturans*' on being antecedent to the segregation into mind and into phenomenal nature," to return to our immersion within corporality as it belongs to the world, nature, *physis*, and "the new concept of intentionality within nature will lead to a new conception of the thought that reflects on it to formulate it."[124] Lingis, in a move unique to his phenomenology, claims that our "sensory-motor organisms respond to the environment as they perceive it," and here we encounter the groundbreaking assertion that it is not just humans that have a unique orientation to the things of the world, things too "have their own orientation."[125] Let us explore the notion of our emotional and *normative* comportment to the world and nature by turning to Harman, whose reading of Lingis provides the rich and fecund ground from which ours takes root and blossoms, for Harman recognizes the monumental importance of Lingis's philosophical move beyond the understanding that *originary ethical* imperatives are always and *only* present to/in the face of the *other*, as we encounter in Levinas. Lingis, as Harman observes, initiates a philosophical move that "never occurs to Levinas,"[126] which is an understanding that the imperatives of the world are inseparable from the "faces" of the *things themselves*, because things too, and not only humans, have faces and show their faces to us, and those faces radiate the imperatives of the world, i.e., "even inanimate objects demand that we treat them in some particular appropriate way."[127] We further develop this interpretation in an even more radical direction in the forthcoming chapters focused on nature, eco-phenomenological ethics, and the *normative* "language" that nature speaks. As stated, intentionality, for Lingis, is always embodied or "corporeal," and this reveals "subjectivity itself as oriented, already turning in the very space it will constitute by constituting objects."[128] The field of kinesthetic sensation or movement-and-perception is in essence an orientation toward the world of sensation that indicates that "the intentionality that means or intends an object is also a movement or orientation with regard to that object,"[129] and this form of corporeal intentionality cannot be reduced to the "performance of a reasoning," and is rather the "kinesthetic awareness of a free variability of orientation."[130] According to Lingis, when we encounter and perceive the outer form of things, "we capture in our postural schema the inner lines of their tensions and orientations" – in the phenomenon where the "corporeal *element* doubles into motor schema and outward aspect"[131] – then, by means of "contracting inner motor diagrams, we quasi-perceive the visible, tangible, and audible form of ourselves turned to them,"[132] in other words, to reiterate what was introduced above, we respond in kind to their unique sensory demands, in terms of a modified phenomenological *homoiosis*. This indicates that the things we encounter in the world are not static presences, mere inanimate objects, that stand before us awaiting our manipulation; rather, their presence holds the supreme power to "lure and threaten us, support and obstruct us, sustain and debilitate us, direct and calm us" – in all, they speak to us and attune us by *enrapturing* us, ultimately moving us to informed action.[133] Thus, just as in Merleau-Ponty, this embodied worldly experience is never in the complete autonomous control of the human, which indicates that the human must be attuned to a world and new reality that we are not in complete control of, that we do not determine in knowledge or otherwise how the world manifests to us, how it uniquely announces its presence with its own inherent and mysterious meanings and hidden values.

In the following description by Lingis we encounter the ontological *theme* – which might also be understood in eco-phenomenological terms – of the human's anticipatory "openness" to nature, which instantiates an attuned moment inspired by the visceral sense of awe and wonder that overtakes us and transforms us in the *face* of nature, facilitating an experience that is undeniably *normative*; and as we read and absorb these powerfully poetic words, let us recall the ground covered regarding Lingis's unique view of phenomenological *intentionality*:

> When I look at the sequoias I do not focus on them by circumscribing their outlines; the width of their towering trunks and the shapes of their sparse leaves drifting in the fog appear as the surfacing into visibility of an inner channel of upward thrust. I sense its force and measure its rise with the movement of my eyes and the upright axis of my body. I comprehend this *uprightness* of their life, not with a concept-generating faculty of my mind, but with the *uprighting aspiration* in my own vertebrate organism.[134]

Here, in this *originary* encounter with the sequoia we do not experience the tree as mere external object that is reproduced internally within consciousness – the temporary unity of sense data that is re-presentational in nature as an *eidetic* phenomenon – and although we have in a sense internalized it, the tree has in turn internalized us. Here, we note that language, for Lingis, occurs authentically only through a listening to the *other*, and so it is possible to state that when nature calls and addresses us, if we are attuned to the call, we offer a rejoinder, and when we listen and respond, we "already know how to pronounce what she says,"[135] and that when speaking to nature, we are capturing the "audible patters" that are spoken to us on our own "motor diagrams."[136] In Lingis's emotive, uplifting, and thought-provoking description, as Harman recognizes, "even the sequoia, that mass of semi-alive organic matter, presents a face to those who encounter it,"[137] and this face and look confronts us and moves us so that "amidst the elemental chaos of the forest and its iridescent gloom and infernal insect chants, we encounter something like a tree-effect," i.e., in a moment of *ek-static* attunement, we intuit or "sense that familiar sequoia thickening grandeur."[138] It is dwelling in the presence of the "uprightness" of the sequoia that inspires the sympathetic – as in the Greek *sympatheo* "*to feel with* or together … sympathize *in a thing*"[139] – "uprighting aspiration" in us and, as stated, in this intimate communion there is an encounter with the *normative*, for there is a form of attunement occurring in the grip of the sublime *grandeur* of nature, namely, we experience and are enraptured by the *tree-effect*, and as Harman contends, to be open and responsive to the "objects populating the earth, and to the elemental medium that supports them, is to enter into the seductive chant of insects, the realm of solar expenditure and vegetative sexuality."[140] As we discuss in the forthcoming chapters, when we inhabit nature we are always already immersed in and poised for an encounter with the ethical imperatives emanating from the *things themselves* that might inspire a radical transformation to our mindset, attitude, and comportment toward the world. For the sequoia as thus described, as Harman observes, "cutting across the ether, turning toward the sun, sucking juice from the soil," always already lives in and inhabits "the domain of the imperative."[141]

In a recent interview, published in "Alterity after Infinity," Lingis categorizes his practice of philosophy *as* phenomenology in the following terms, stressing that even in the most minimal sense, and here recall the above description of the great sequoias, "all insights in philosophy come through describing something, perhaps something that has been described again and again," but what good philosophy does and accomplishes is to describe it "more closely, more accurately," and beyond this, to attempt to describe "something in a phenomenon that had not been taken account of."[142] Randolph Wheeler calls Lingis's method an *innovative* mode of phenomenological description, through which he "reverses the usual order of philosophical elaboration that proceeds from concepts to details," and this is because he begins with "vibrant details that illustrate his ideas," an approach which, as related to our concerns, "situates his method squarely in the 'really existing' world, in which his explication of concepts enhances our understanding of the world, and re-enhances the world."[143] We get this sense of vibrant and rich description in and through his responding to the call of language that is poetic in nature. For in the practice of what Lingis originally labeled, in a study focused on expanding the thoughts of both Husserl and Merleau-Ponty, *interrogative ontology*, language is the medium that teases out the things it describes, intimating while sheltering the mystery inherent within the things drawn out from the silent depths of their dwellings, for language here is not the equivalent of assertions or propositional locutions. For Lingis, language is embraced as a fluid and receptive form of expression and communication through which "occult commerce of metaphors" expresses, or gestures and intimates, "in its lateral movement, its metaphors, its traversal organization, the identifiable style, the structural laws, the regular relationships and latent logic of the things of the world."[144] It is interesting that in the phenomenology, or practice of *lived world research*, of van Manen we encounter a decidedly similar view of language that at once emerges from the *interrogative* mode and is understood as *poetic*, *descriptive*, and *hermeneutic*. "Language," contends van Manen, when it "authentically speaks the world rather than abstractly speaking *of* it is a language that reverberates the world."[145] And, just as in Lingis's phenomenology, van Manen stresses that we "must engage language in terms of a primal incantation or poetizing which harkens back to the silence from which the words emanate."[146] Indeed, concluding this section with Harman's thoughts, this talk of phenomenology and language is very much in line with the way he classifies and describes the general practice of *good philosophy*, which relates directly to our conception of eco-phenomenology, and includes for him "good writing," which is "not just clear and devoid of 'fuzziness,'" but is also "*vivid* writing that brings its subject to life rather than replacing it with bundles of explicit and verifiable qualities,"[147] i.e., allowing what is written about and spoken of to live, to dance freely beyond the strictures of both formal knowledge and literal communication.

Notes

1 Rainer Maria Rilke, *Letters to a Young Poet*, trans. M. D. Herter Norton (New York: W. W. Norton & Company, 2004), 16.
2 *Fourth National Climate Assessment*, http://science2017.globalchange.gov.
3 David Wood, "What Is Eco-Phenomenology?" in *Eco-Phenomenology: Back to the Earth Itself*, eds. Charles Brown and Ted Toadvine (Albany: SUNT Press, 2003), 232.
4 Ibid.
5 Arne Naess, "The Shallow and the Deep, Long-Range Ecology Movement: A Summary," *Inquiry*, 16, no. 1 (1986): 8.
6 Michael Zimmerman, "Heidegger, Buddhism, and Deep Ecology," in *Cambridge Companion to Heidegger*, ed. Charles Guignon (Cambridge: Cambridge University Press, 1990), 242.
7 Arne Naess, "The Deep Ecological Movement: Some Philosophical Aspects," in *Environmental Philosophy: From Animal Rights to Radical Ecology*, eds. Michael Zimmerman, et al. (Englewood Cliffs: Prentice Hall, 1993), 99.
8 Ibid.
9 Ibid., 200.
10 Ibid., 93.
11 Ibid., 96.
12 Ibid., 209.
13 Ibid.
14 Ibid., 104.
15 Ibid., 200.
16 Ibid., 196.
17 Ibid., 203.
18 Ibid.
19 Ibid., 197.
20 Wood, "What is Eco-Phenomenology?" 224.
21 Louise Economides, "Romanticism and Buddhism: Blake, Heidegger, Buddhism, and Deep Ecology," *Romantic Circles* (2007): www.rc.umd.edu, 1–2.
22 Ibid., 2.
23 Dan Zahavi, "The Phenomenological Tradition," in *Routledge Companion to Twentieth Century Philosophy*, ed. D. Moran (London: Routledge, 2008), 10.
24 David Detmer, *Phenomenology Explained: From Experience to Insight* (Chicago: Open Court, 2013), 1.
25 Zahavi, "Phenomenological Tradition," 8.
26 Ibid., 6.
27 Ibid., 8.
28 Ibid., 13.
29 Ibid., 13.
30 Ibid.
31 Ibid., 9.
32 Max van Manen, *Researching Lived Experience* (Albany: SUNY Press, 1990).
33 Karl Jaspers, *Way to Wisdom*, trans. Ralph Manheim (New Haven: Yale University Press, 1954).
34 Daniel Dahlstrom, "Introduction," in *Philosophy of Mind and Phenomenology: Conceptual and Empirical Approaches*, eds. Daniel Dahlstrom, et al. (London: Routledge, 2016), 1.
35 Zahavi, "Phenomenological Tradition," 2
36 Martin Heidegger, *The History of the Concept of Time: Prolegomena*, trans. Theodore Kisiel (Bloomington: Indiana University Press, 1992).
37 Ibid., 50–51.

38 Mark Wrathall and Hubert Dreyfus, "A Brief Introduction to Phenomenology and Existentialism," in *A Companion to Phenomenology and Existentialism*, eds. Mark Wrathall and Hubert Dreyfus (Malden: Wiley-Blackwell, 2008), 9.
39 Ibid., 5.
40 Detmer, *Phenomenology Explained*, 18.
41 Ibid.
42 Graham Harman, *Towards Speculative Realism* (New York: Zone Books, 2009), 105.
43 Graham Harman, *Immaterialism: Objects and Social Theory* (Cambridge: Polity, 2016), 8.
44 Ibid., 8–12.
45 Harman, *Towards Speculative Realism*, 112.
46 Ibid., 113.
47 Ibid.
48 Ibid.
49 Graham Harman, "The Future of Continental Realism: Heidegger's Fourfold," *Chiasma*, 3, no. 1 (2017): 3.
50 Sean Esbjorn-Hargens and Michael Zimmerman, *Integral Ecology: Uniting Multiple Perspectives in the Natural World* (Boston: Integral Books, 2009), 507.
51 Charles Brown and Ted Toadvine, "Eco-Phenomenology: An Introduction," in *Eco-Phenomenology: Back to the Earth Itself*, eds. Charles Brown and Ted Toadvine (Albany: SUNY Press, 2003), x.
52 Ibid., xii.
53 Monika Langer, "Nietzsche, Heidegger, and Merleau-Ponty: Some of Their Contributions and Limitations for Environmentalism," in *Eco-Phenomenology: Back to the Earth Itself*, eds. Charles Brown and Ted Toadvine (Albany: SUNY Press, 2003), 106.
54 Brown and Toadvine, "Eco-Phenomenology: An Introduction," xi.
55 Ibid., xii.
56 Iain Thomson, "Ontology and Ethics at the Intersection of Phenomenology and Environmental Philosophy," *Inquiry*, 47, no. 1 (2004): 381–382.
57 Ibid., 399.
58 Brown and Toadvine, "Eco-Phenomenology: An Introduction," x.
59 Wood, "What is Eco-Phenomenology?" 231.
60 Ibid.
61 Ibid.
62 Ibid.
63 John Llewelyn, "Prolegomena to Any Future Phenomenological Ecology," in *Eco-Phenomenology: Back to the Earth Itself*, eds. Charles Brown and Ted Toadvine (Albany: SUNY Press, 2003), 58.
64 Ibid., 53.
65 Hans-Georg Gadamer, *Truth and Method*, trans. Joel Weinsheimer and Donald Marshall (London: Continuum, 1989), 19.
66 Graham Harman, *Object-Oriented Ontology: A New Theory of Everything* (London: Penguin, 2018), 7.
67 Harman, *Immaterialism*, 32.
68 Ibid.
69 Harman, *Object-Oriented Ontology*, 186.
70 Llewelyn, "Prolegomena," 54.
71 Ibid., 56.
72 Richard Polt, *The Emergency of Being: On Heidegger's Contributions to Philosophy* (Ithaca: Cornell University Press, 2006), 159.
73 Llewelyn, "Prolegomena," 59.
74 Maurice Merleau-Ponty, *Phenomenology of Perception*, trans. C. Smith (London: Routledge, 1989), vii.

75 van Manen, *Researching Lived Experience*, 177.
76 Martin Heidegger, *Introduction to Metaphysics*, trans. Gregory Fried and Richard Polt (New Haven: Yale University Press, 2000).
77 Llewelyn, "Prologomena," 59.
78 van Manen, *Researching Lived Experience*, 39–40.
79 Ibid., 101.
80 Ibid.
81 Ibid., 79.
82 Michael Steinmann, "Phenomenological Perspectivism," *Heidegger Circle Archive* (45th Annual Meeting), 48.
83 van Manen, *Researching Lived Experience*, 88.
84 Alphonso Lingis, *Foreign Bodies* (London: Routledge, 1994), 11.
85 Ibid.
86 Brown and Toadvine, "Eco-Phenomenology: An Introduction," xi.
87 Eric Matthews, *The Philosophy of Merleau-Ponty* (Ithaca: McGill-Queen's University Press, 2002), 34.
88 Brown and Toadvine, "Eco-Phenomenology: An Introduction," xii.
89 Lingis, *Foreign Bodies*, 32.
90 Alphonso Lingis, *Phenomenological Explanations* (Boston: Martinus Nijhoff Publishers, 1986), 45.
91 Ibid., 49.
92 Ibid.
93 Ibid., 59.
94 Ibid., 52.
95 Ibid., 53.
96 Vincent Descombes, *Modern French Philosophy*, trans. L. Scott-Fox and J. M. Harding (Cambridge: Cambridge University Press, 1980), 56.
97 Maurice Merleau-Ponty, *Prose of the World*, trans. John O'Neill (Evanston: Northwestern University Press, 1964), 238.
98 Merleau-Ponty, *Prose of the World*, 248.
99 Langer, "Nietzsche, Heidegger, and Merleau-Ponty," 115.
100 Matthews, *The Philosophy of Merleau-Ponty*, 29.
101 Ibid., 35.
102 Ibid., 39.
103 Ibid., 35.
104 Ibid., 38.
105 Merleau-Ponty, *Phenomenology of Perception*, 440.
106 Ibid.
107 Matthews, *The Philosophy of Merleau-Ponty*, 43.
108 Ibid.
109 Lingis, *Phenomenological Explanations*, 33.
110 Ibid., 34.
111 Ibid., 33.
112 Ibid., 34.
113 Ibid., 33.
114 Ibid., 35.
115 James Magrini, "Thinking Through Nihilism: The Reclamation of Embodied Thought for Enhancing Cultural Practices," *Analysis and Metaphysics*, 12, no. 1 (2013).
116 F. E. Peters, *Greek Philosophical Terms: A Historical Lexicon* (New York: New York University Press, 1967), 87.
117 Alphonso Lingis, *The Imperative* (Bloomington: Indiana University Press, 1998), 70.
118 Ibid., 63.
119 Ibid.

120 Ibid., 59.
121 Ibid., 61–62.
122 Ibid., 62.
123 Ibid.
124 Ibid., 37.
125 Lingis, *Foreign Bodies*, 3.
126 Harman, *Towards Speculative Realism*, 18.
127 Harman, *Object-Oriented Ontology*, 67.
128 Lingis, *Phenomenological Explanations*, 25.
129 Ibid., 24.
130 Ibid., 25.
131 Ibid.
132 Ibid., 16–17.
133 Lingis, *Foreign Bodies*, 42.
134 Ibid., 17.
135 Ibid.
136 Ibid.
137 Harman, *Towards Speculative Realism*, 17.
138 Ibid., 18.
139 Liddell and Scott, *A Lexicon: Liddell & Scott's Greek–English Lexicon* (Mansfield Centre: Martino Publishing, 2015), 664.
140 Harman, *Towards Speculative Realism*, 20.
141 Ibid., 21.
142 Alphonso Lingis, "Alterity after Infinity: Interview with Alphonso Lingis (2015)," in *Passion in Philosophy: Essays in Honor of Alphonso Lingis*, ed. Ralph Wheeler (New York: Lexington Books, 2017), 103.
143 Ralph Wheeler, "Imperative Innovations," in *Passion in Philosophy: Essays in Honor of Alphonso Lingis*, ed. Ralph Wheeler (New York: Lexington Books, 2017), 119.
144 Lingis, *Phenomenological Explanations*, 53.
145 van Manen, *Researching Lived Experience*, 13.
146 Ibid.
147 Harman, *Immaterialism*, 32.

2 The cultural *Weltanschauung*

"Naturalistic" framework and our interaction with nature

Introducing the "Cultural Framework"

We understand the "Cultural Framework" (CF) in terms of an epochal conceptual lens through which we reveal, understand, and discourse about the world we inhabit; it tacitly "colors" our interpretations and behaviors, influencing the unfolding of our entire Being-in-the-world as *dwelling-with-nature*. In eco-phenomenology, the notion of CF is primarily traceable to Heidegger and Husserl, this despite the fact that each philosopher held a differing view and practice of phenomenology. In Husserl, if we examine both early and later manifestations of his thought, it is possible to understand the CF in terms of the dominance of *naturalism/scientism* and the influence it has on the *natural attitude* and indeed the entirety of the human's "life-world" (*Lebenswelt*). This indicates that the human's world is dominated by an overarching scientific-technological attitude, which for Husserl, represents the *crisis in the European sciences*. Looking to Heidegger, it is possible to understand the CF in terms of the overarching mood (*Stimmung*) or mode of attunement associated with *das Ge-stell*. Michael Bonnett argues that this idea of the CF determines

> what is visible and what remains invisible; what is relevant and important and what is not; what is a problem and what is not; what counts as a valid way of proceeding in addressing a problem and what does not.[1]

Bonnett recognizes that it is the "complex and intimate interplay between cultural motives and artifacts and the otherness of nature that constitutes world-formation," i.e., the "ever evolving and historically grown overarching form of sensibility through which we engage the world has arisen in fundamental respects in response to nature."[2] This framework, as related to eco-phenomenology, might also be understood as a climate of opinion, cultural worldview (*Weltanschauung*), ideological framing, or, as stated above, conceptual lens. In the philosophy of Husserl, Heidegger, and later postmodern philosophy (e.g., Derrida), it is not uncommon for the CF to be expressed in the following terms expressing a univocal meaning: *Naturalism*; *Framework of Scientism*; *Metaphysics of Presence*; *Metaphysical Instrumentalism*; *Metaphysics of Mastery*; *Voluntaristic Metaphysics*; *Productionist Metaphysics*; *das Ge-stell*; *En-Framing*; and the like.

The influence of the CF works in a tacit manner, but might be made explicit (wrested from concealment) through the practice of phenomenology/eco-phenomenology when examining and interpreting *praxis*, specifically focused on the ways in which we interact with nature in the original "embodied" mode of *immersed*-intentional comportment. We are also concerned with how this interpretation is expressed within various modes of worldly involvement, including the personal level, communal level, governmental level, and the like. The influence of the CF is comprehensive and grand in scope, and as Lester Embree argues, includes the cultural science of the environment, aesthetic-recreational or valuational environmentalism, and practical-political types of environmental encounters.[3] Our task in this chapter is to explore two problems that this metaphysical CF engenders, namely:

1 The CF limits the way we can speak about/of the world – e.g., Gabriel Marcel claims that our encounters with nature are *sublime* and *mysterious* in terms of an indefinable and in many ways ineffable *presence*. Here the language we turn to is non-literal and more poetic in tone because these types of encounters (with others, nature, works of art), while undeniably palpable, are expressive of ideas that are "fundamentally incommunicable, [to the extent that] we cannot even communicate [their] pure essence[s] to ourselves";[4]
2 The CF restricts our view of nature to a mechanical causal nexus or "store" of non-human resources open for the taking. If nature is "objectified," in terms of its *instrumental* and *economic* value for human profit and consumption, it is reduced to a product of our *anthropocentric* categorizations, narratives, texts, and uses. This view is antithetical to the eco-phenomenological view that nature has its own "intrinsic" value outside of any instrumental value, or otherwise, the human places on it. The CF forecloses the potential for the eco-phenomenological move to legitimately consider a *normative* approach to our interaction with nature, grounded in certain "ethical" imperatives that are inherent within and "intimated" by nature itself.

We note that for Husserl and Heidegger, the CF is a thoroughly Western phenomenon, traceable, in the former, to the rise of the European sciences and, in the latter, to the "first beginning" of Western Metaphysics and Plato's misunderstanding of truth as *originary* disclosure (*aletheia*), a phenomenon which Heidegger claims was intimated by the pre-Socratic philosophers. However, we claim that the CF, despite displaying blatant characteristics of both Cartesian and Kantian metaphysics and epistemology, transcends the Western perspective and enacts a more grand and sweeping influence. For, we argue, it manifests at a global level wherever there is the insidious influence of *naturalism/scientism*, be that in Africa, the Americas, Europe, Asia, or Oceania. To provide but one example among innumerous instances, consider the heinous environmental crime of "finning" sharks for the sale of their fins and tails in the foreign restaurant trade for use in *shark fin soup*. This practice involves capturing mass numbers of sharks and de-finning them while they are alive and then throwing them back into the ocean to inevitably die slow and agonizing deaths. Their severed fins and tails garner upwards of €600 in the Asian market. Here, we encounter an exploitive industrialized and systematized practice that is wholly *anthropocentric* in nature, driven exclusively by *economic* aspirations, which tortures and kills wildlife while simultaneously disrupting the ocean's food chain that depends on the predation of sharks that are killed by foreign fisherman.[5]

The CF as we are conceiving it, drawing on the philosophy of John Gray, is in great part traceable to the human's *deification* of reason and the privileging of an *instrumental* view of science and technology, which are applied and exercised with ruthless efficiency serving anthropocentric ends despite the environmental destruction left in their wake. In this pernicious view, according to Gray, which eco-phenomenology seeks to transform, "humans are unique in that, using their power over nature given them by science, they can create a world better than any that has existed before," and so "the earth is simply a mass of resources for human use, and the other animals with which we share it have no value in themselves."[6] Esbjorn-Hargens and Zimmerman echo these thoughts in their observation that the technological move to dominate nature is a thoroughly modern phenomenon. In such a milieu, given structure by the dominant CF – as *metaphysics of mastery* – overcoming of this destructive drive to heedlessly exploit nature is possible, not through new technological advances or changes to the social order, but rather through the far more difficult project and occurrence of facilitating a "healthy expression of each level of [human development]," and this includes encouraging the human's "interior development in many different capacities"[7] that might potentially be reflected in and expressive of a re-conceptualized relationship to nature, which we have stated in Chapter 1 requires a radical transformation to our worldview, belief system, or *attunement*. In the next two sections we move to further explore crucial aspects of both Husserl's and Heidegger's phenomenology as related to our elucidation and critique of the CF.

The *Natural Attitude* and the *Lebenswelt* in Husserl and eco-phenomenology

The rise of naturalism-scientism and the Cultural Framework

As introduced in Chapter 1, when eco-phenomenological studies draw from Heidegger's philosophy they are primarily focused on a form of *attunement* (*Stimmung*) – *Metaphysics of Presence/Mastery* – that must be overcome or transformed as related directly to *das Ge-stell*. However, when eco-phenomenology takes its lead from the philosophy of either Edmund Husserl or Maurice Merleau-Ponty, the concern is transforming our attitudes by focusing on revealing and critiquing the *natural attitude* as it has been permeated and, in the extreme, dominated by *naturalism* or *scientism*, which represents one mode of world-disclosure comprising what in the later phenomenology of Husserl is called the *Lebenswelt* or "life world."[8] Eco-phenomenology identifies the danger that when entrenched in the *natural attitude* that is dominated by *objectivism*, *naturalism*, or *scientism*, we see and experience reality solely from the perspective of an attitude that resembles a "natural scientist, as abstracted from subjective activities," and in this attitude human rationality (reason) – which is also expressed through *calculative* thought/formable knowledge – is privileged above other modes of world-disclosure and reduced to the realms of the mathematical, mechanical, and causal. This epistemic prejudice forecloses our legitimate engagement with other modes of world-disclosure that are related to both *ethics* and *aesthetics*, which of course are expressive of *normative* or *axiological* concerns related to value and meaning, and according to eco-phenomenology, these ways of "knowing" the world might be enhanced through a phenomenological return to the original and pre-reflective manner in which we inhabit and "constitute our life-world."[9] As we learn in the next section, such movements as *secular* or *neo-liberal humanism* are expressions of the *metaphysics of presence* because they are instantiated by a radical form of *naturalism*. For example, in Gray's philosophy, what we have identified as the CF is linked indelibly to *secular humanism* and the unrealistic and dangerous – and even "pseudo-religious" – belief in unadulterated *human progress*, which is prevalent in the world wherever there is advanced technological industrialization.[10]

With this understanding we explore both the *natural attitude* and "life world," because the issue of how the CF is at work shaping and in great part determining our relationship to nature, along with the concern for the vast scope of its influence, are issues more complex than they appear at first blush. Thus, we turn to explore the various layers composing the human's overall comportment based on its understanding of the world, by bringing together concepts of early and later Husserl, which might be understood in tripartite:

1 the phenomenological mode of the human's *originary* and pre-reflective involvement – *immersed*-intentional comportment – specifically, for our purposes, with the value-laden and meaning-invoking world of nature;
2 the overarching "lived world" (*Lebenswelt*) of our comportment, and
3 the *natural attitude* that has the potential to influence via *naturalism* or *scientism* (2) and sets the human being at an ever greater (ontological) distance from the mode of existence expressed in (1).

In relation to this foregoing three-part division, as eco-phenomenologist Erazin Kohak contends, it is not about inventing "alternative realities," but is rather about pointing, by means of phenomenological revelation and analysis, to "alternative ways of organizing or constituting the one reality ["life world"] amid which we live, breathe, and have our being."[11] The *natural attitude* is a "commonsense" mode of world-disclosure or context of revelation, which is always already permeated by epistemological and metaphysical assumptions that are linked to the views of both the natural and the technological sciences. Husserl states that in the *natural attitude*, the thinking and perceiving subject knows, and beyond,

> is certain that the surrounding world that is posited as existing in the manner of immediate intuition is only the intuited piece of a total surrounding world and that things continue on further in endless space ... the subject knows also, or is certain ... that the world is and that it, the experiencing subject itself, is in the middle of this world.[12]

So, through the lens of the *natural attitude*, in its grip or under its sway, we impose upon the world – including the realm of pre-theoretical *immersed-*intentional comportment – in an unknowing, passive, and discreet manner, a schema for revelation and interpretation grounded in the reductive understanding of the world – and by extension nature – in terms of a mechanistic phenomenon, unfolding according to strict laws given structure by antecedent cause-and-effect. Here, we embrace the existence of a mind-independent reality that is given life, form, and validation through the imposition of our epistemological categorizations and structures, and this view is expressive of the Cartesian subject/object and internal/external divide. Our immersion within the *natural attitude* as a way of disclosing the world, according to Shaun Gallagher, serves as motivation for phenomenologists seeking to gain a "critical perspective on the way the natural attitude works,"[13] and so, as discussed in Chapter 1, there is an attempt made to hold it in abeyance in order to gain access or open a vista into a new way of seeing and experiencing the pre-reflective realm of *immersed-*intentional comportment. This, however, as Zahavi points out, is not to deny the *natural attitude*, for in Husserl the *epoche* and *eidetic* reduction are not practiced to "doubt, neglect, abandon, or exclude reality from our research," but rather their purpose is to suspend or momentarily neutralize "a certain dogmatic *attitude* toward reality, thereby allowing us to focus more narrowly and directly on reality just as it is given."[14] By suspending this attitude and then "thematizing the fact that reality is always revealed and examined from some perspective or another, reality is not lost from sight," but is made explicit and accessible for a legitimate form of investigation.[15] Indeed, according to Gallagher, our attempts to "highlight pre-scientific perception can consequently be interpreted as an implicit criticism of scientism."[16] To explore *pre-scientific perception* – to reiterate what was introduced in Chapter 1 – we put the *natural attitude* in phenomenological "brackets." To do this is to at once gain access into a new way of experiencing the phenomena with which we interact in meaningful ways, as we bring into focus, by means of a juxtaposition, the structure of the *natural attitude* that has seemingly been layered over the so-called phenomenological or *originary* mode of worldly dwelling, and in this way open it up for analysis and critique. For as Zahavi stresses, philosophy must question the assumptions we hold in the *natural attitude*, and this questioning and inquiry should be "critical and undogmatic, shunning metaphysical and scientific prejudices" as it is "guided by what is actually given, rather than by what we expect to find given our theoretical commitments."[17] It is imperative, however, to recall from Chapter 1 that the goal of phenomenology is to potentially return from our pre-theoretical reflections enlightened, in ways that might transform our theoretical views of the world that were initially and necessarily "bracketed."

As stated, there is a way that avoids conflict, of merging early and later Husserl to understand the *natural attitude* and the *Lebenswelt* in terms that reveal the primacy of *naturalism* (*scientism*) at work influencing the *natural attitude* and beyond, the human being's entire *Lebenswelt*. According to Husserl, the "life world," whose structures can be studied in a phenomenological manner, is not a world wherein we encounter physical objects in space; rather it consists of a meaningful interplay of relations within overlapping and interconnected contexts within which we interact and share with others, i.e., "each 'I-the-man' and all of us together," as Husserl argues, "belong to the world as living with one another in the world; and the world is our world, valid for consciousness as existing precisely through this 'living together.'"[18] Kohak, in his vision of eco-phenomenology, stresses the important point that the *Lebenswelt* "*is neither value free nor meaningless*," for it is a world of purposeful and meaningful activity, but it also a world that "not only *is*, but from the beginning also aids or hinders the activity that constitutes it."[19] The way we inhabit the *Lebenswelt* – the things we do, the goals we work to achieve, the values we adopt, hold, and pass along, the ways we interact with others and nature – is influenced by the varying attitudes, climates of opinion, and in short, the ways in which we are attuned. Eco-phenomenology claims that not only is our *natural attitude*, and its mode of disclosure, dominated by a scientific and technological view, beyond this, *naturalism* or *scientism* permeates and infects the entire *Lebenswelt*. For example, when we interact with nature, the world comes to presence for us in and through a CF that is reducible to, because it is structured by, rote economic accounting, the anthropocentric drive toward instrumentality, and the misguided faith in unbounded scientific progress. This represented a "crisis" for Husserl, where abstractions and conceptual theoretical assumptions are privileged and accepted as representing "reality" *in toto*, which has the inevitable effect of obscuring the fact that it is through our original, immersed, and pre-reflective experiences of, and in, the world that all scientific views initially acquire their meaning.[20] Through the experience of what it is possible to label *ontological obfuscation*, we have lost sight of the fact that it is the realm of pre-reflective *immersed*-intentional comportment that actually constitutes the historical and systematic foundation of the sciences, as Gallagher expresses in the following manner:

> Scientific discourse is embedded in the world of experience, in the experiential world, and if we wish to comprehend the performance and limits of science, we have to investigate the original experience of the world of which science is a higher-order articulation.[21]

More will be said about *scientism* below, but for now we shall briefly explore the concept in order to justify the critical project of eco-phenomenology, which as introduced, is both a *diagnostic* and *prescriptive* practice and endeavor. When *naturalism* or *scientism* dominates the *natural attitude* and the "life-world," according to Brown and Toadvine, environmental destruction and crises are inevitable, brought on by the change to "core beliefs within our worldview that sanction, legitimate, and even encourage the domination and technological control of nature."[22] Bonnett defines *scientism* as the modern phenomenon of believing that "experimental scientific thinking has a privileged access to the nature of reality" in which scientific "methods, findings and constructions reveal what is 'really' real and that therefore it can assume the mantle of arbiter for thinking in general."[23] As noted by Gallagher, there is a certain blind spot in the sciences that precludes them from grasping their origins, and this is because science is "so absorbed in its investigation of the natural (or social/cultural) world that is doesn't pause to reflect upon its own presuppositions and conditions of possibility."[24] For example, the empirical sciences operate on the naïve, but necessary, assumption that there is a mind-independent reality that can be investigated in an objective and value-neutral manner and, ultimately, that scientific investigations hold the power to "provide strict and objectively valid knowledge about this given realm."[25] To reiterate, *naturalism* is not a mode of disclosure linked directly to the technical sciences, but is a view in which the presuppositions of the sciences insidiously creep their way into our everyday ways of disclosing the world, as expressed in the *natural attitude*. This phenomenon must be contrasted and questioned by a philosophical attitude that seeks to understand the foundation of scientific thought and indeed all experience. However, as Gallagher observes, a strict *naturalism* or *scientism* in the manner of positivism "denies the existence of a unique philosophical method, and claims that philosophers should consider their own work to be directly continuous with the natural scientist."[26] Philosophy, however, as we find in the practice of phenomenology and eco-phenomenology, cannot and so does not contribute to the type of positive empirical knowledge associated with the sciences, but rather "investigates the basis of this knowledge and asks how it is possible,"[27] and this is an issue we explore further in Chapter 3 focused on Harman's *Object-Oriented Ontology*. With this in mind, the task of phenomenology is to reveal for analysis, i.e., make explicit what has hitherto remained tacit or concealed, the three modes of world-disclosure introduced above: (1) The "phenomenological" pre-theoretical world of *immersed*-intentional comportment, (2) The *Lebenswelt*, and (3) The *natural attitude* – with the goal of uncovering and identifying to what extent and degree positivistic or scientific thought dominates our *originary* experience of the world and Nature. Ultimately, eco-phenomenology seeks to offer a potential way in which we can overcome or transcend the grip of the dominant CF that is given structure by *naturalism* or *scientism*; and this move can be related to Husserl and Merleau-Ponty and, to a lesser extent, Heidegger, considering the latter's decidedly pessimistic view of the possibility of humanity transcending attunement of modern technology (*das Ge-stell*).

The Cultural Framework as *das Ge-Stell*

Heidegger's "epochs of Being/Beyng" in relation to nature

Another manifestation of the CF in eco-phenomenology is traceable to Heidegger, and for the most part this marks a turn to the 1954/1977 essay "The Question Concerning Technology," which famously elucidates the oppressive "en-framing" effect of *das Ge-stell*.[28] Although this form of attunement instantiates for Heidegger the so-called "essence of technology," it is not itself technological in nature. *Das Ge-stell* instantiates a specifically modern mode of revealing truth, which is simultaneously for Heidegger a mood (*Stimmung*) – or mode of *attunement* – through which truth reveals itself, i.e., this mood *colors* the manner in which all things (others, world, nature, and so forth) show up for the human being. We note that Heidegger claims "moods" are more primordial and so antecedent to any psychological structures, and so it is possible to state that moods for Heidegger represent "first-order" modes of world-*disclosure* while psychological and epistemological categories represent "second-order" modes of *disclosure*. Although we do not embrace a "transcendental phenomenological"[29] (Heideggerian) reading of nature, which would require an analysis of various *Stimmungen*, it will prove beneficial to briefly explore the notion of *das Ge-stell* in Heidegger. This will lead into a discussion of Heidegger's theory regarding the history of Being's *disclosure* – or, as "Later" Heidegger prefers, "Beyng" – within which the modern age is but one among many ways that Being comes and has come to *presence* as a historical defining force, in terms of opening and grounding *historical epochs*, highlighting, for our purposes, the human's relationship with nature (the Earth) that is unique to the "cultural" encounter with Being's *presencing* as humans orient toward meanings that are destined, revealed, and made available – given by Being – for their appropriation in a historical event that Heidegger refers to as *Ereignis*, or what we might say is the *authentic appropriation of what is most historically appropriate for the human being*.

Das Ge-stell is a framework (*Gestalt*) or mode of "setting-up" linked to a restrictive mode of world-*disclosure* ("calculative thought") that conceals, according to Heidegger, the essence of the truth of Being. In the grip of technology's *Ge-stell*, the world shows up in ways consistent with an "instrumental" view of nature, and hence the human's relationship with nature is reducible to the anthropocentric, instrumental, economic, and scientific. As opposed to the fundamental attunement of great works of historical-founding art (e.g., Greek temples and medieval cathedrals), the "en-framing" effect of *das Ge-stell* reduces all things to commodities and *mere* objects that are usable and then disposable: Indeed, the entire world shows up in terms of what Heidegger identifies as "standing reserve" (*Bestand*), which indicates that nature serves, in no uncertain terms, as an *energy* source, indicating that energy/resources are extracted from the natural world and stored up (as *standing reserve*) for future use. In the thralls of *technology*, we are attuned to quantify our existence in terms of locating, extracting, and using or exploiting the earth's resources for our technological-economic advancement, because in the mood of *das Ge-stell* the privileged mode of world-*disclosure* highlighting the modern epoch, as opposed to the "meditative" thought associated with authentic *thinking* (and *not* philosophy), is "calculative thought,"[30] thought that is set against objects to "know" them with cold mathematical precision and objectivity.

Here, we encounter a thoroughly modern view of the human being, nature and, ultimately, the manner in which humans interact with nature. To draw but two examples from Heidegger in relation to what we have introduced above, under the sway of technology's attunement, mountains fail to "show up" or reveal themselves *as* mountains, they instead *presence* in terms of potential mineral stores for mining or fossil fuel sources for drilling. Similarly, the river, no longer coming to *presence* as river *qua* river, is reduced to a dynamic force that might be harnessed for use as a form of hydro-electric energy, its beauty and intimate and "emplaced" connection to its surrounding geography and topology, its role as home to water fowl, fish, and insects, as a thoroughfare for travelers, be damned.[31] Importantly, for Heidegger, what is lost in such instances where humans are attuned by *das Ge-stell* is the mindset that fosters and facilitates the human's *releasement* (*Gelassenheit zu den Dingen*) over to things and the openness to the *mystery* (*Offenheit für das Geheimnis*) in all things, which ultimately attunes us to the "letting be" of the Earth *as* Earth. As Zimmerman stresses, "One 'lets things be' not for any external goal, but instead simply from a profound sense of identification with all things."[32] In this view of nature shaped by *das Ge-stell*, as stated above, the human no longer shares a relationship with nature that could be described as, if we look to Arne Naess and Deep Ecology, "symbiotic" or "holistic" – a "view that decenters humankind and emphasizes care for all beings"[33] – and rather it is a relationship enacted through a *meaning-making* schema structured according to metaphysical, hierarchical polar oppositions, where the human being assumes the disingenuous and inauthentic role of nature's supreme arbiter and master, which opens the door for nature's destruction through the human's insatiable consumption and exploitation through the forces of *machination* and *violence*.[34]

In the essay, "The Origin of the Work of Art," when discussing the power of art as a historical-destining- *founding* that is at once a "beginning" (*Ursprung*), Heidegger claims that "when beings as a whole, as beings themselves, demand a grounding in openness, art attains to its historical essence as foundation."[35] The foundation occurred for the first time in Greece and then later "the realm of beings thus opened was then transformed into a being in the sense of God's creation," which happened in the Middle Ages, and here we might imagine the experience of the Earth and the holy, occurring in the communal Christian religious experience of the medieval cathedral and no longer in the ancient Hellenic temple. The type of founding associated with art occurred at least two times in history; however, the modern age has been lost to the cultural-historical founding work of art, and so is an age expressive of nihilism and the rise of the *will to power*, "beings became objects that could be controlled and penetrated by calculation."[36] Although Heidegger limits his discussion of Being's historical founding in terms of three historical occurrences, we explore five such moments when the truth of Being opened world epochs. The historical essence of these historical epochs is obscured, and we require what Heidegger terms a "dismantling" (*Destruktion*) in order to lay bare the essence. Heidegger states in the essay "Time and Being," "What is historical in the history of Being is determined by what is sent forth in destining ... we speak of the epochs of the destiny of Being" and, further, that what is appropriate shows itself as destiny and in the "belonging together of the epochs. The epochs overlap each other in their sequence so that the original sending of Being as presence is more and more obscured in different ways."[37]

The epochs we shall briefly discuss instantiate specific and historical manifestations of one or another CF as we have introduced the idea. Thus, the Heideggerian notion of CF is not a contemporary phenomenon; rather, the manifestation of multiple cultural frameworks represents what Heidegger understands as the "hidden history of the West," which, as stated above, reveals the manner in which "worlds" manifest as "epochs of Being." The *History of Being*, for Heidegger, is actually the history of *productionist metaphysics*, and this history, as Zimmerman brings to our attention, lives on three interrelated fronts: "(1) the change of being from *energeia* to actuality or reality, (2) the change of truth from *aletheia* to certainty, and (3) the change of substance from the change of *hypokeimenon* to *subjectum*."[38] We must note that Heidegger had many versions of this story or history, "none of which are completely consistent," as Zimmerman points out.[39] Indeed, our retelling of the *History of Being* draws on different Heideggerian sources than Zimmerman's recounting, for we are specifically concerned with Nature's role in this history. In light of Heidegger's claim, and true to our understanding of phenomenology and its project, we reiterate that although the CF we are analyzing in relation to our eco-phenomenological analysis bespeaks and indeed determines a relationship to self, others, world, and nature, this relationship is hidden from explicit view, for we are socialized into this "world view" – or mode of *attunement* – and often do not question its existence or origin, and so the revelation of the existence of the CF is an intimate concern for the project of phenomenology. The historical epochs are identified as follows, with the "modern" epoch being of the greatest concern for us, and we note that each of the five CFs, which have corresponding modes of *attunement* (*Stimmungen*), open a vista into the historical ways that humans have understood and interacted with nature, and continue to do so.

The Homeric (Archaic) Greek epochal framework

"*Φυσις* is the determination of being found at the commencement,"[40] where *physis* is understood in terms of the revelation of truth as *aletheia*. This age instantiates the human's relationship with "*physis*" or nature highlighted by the sublime and reverential sense of wonder (*thauma*) the Greeks experienced in the presence of the natural world opening up and revealing itself as all things rose up, grew, and developed and then eventually degenerated and withered away. *Physis* is an unaided (by the human) form of "bringing forth" (*Her-vor-bringen*) and so is an *originary* and supreme form of *poiesis*, as the "arising of something out of itself," and what "presences by means of *physis* has the bursting open belonging to the bringing-forth, e.g., the bursting of a blossom into bloom, in itself (*en heauton*)."[41] To this end, this is the opening up and revelation of the natural world in which the human participates or experiences, but does not aid in the process, as would an artisan or craftsman (*technites*). This is also an age when the Greeks were attuned to the "fire from heavens," which revealed for the Homeric Greeks their proper vocation and historical destiny for appropriation as epochal event (*das Ereignis*).[42]

The Classical Greek epochal framework

This historical age is highlighted by a relationship between the human being and nature as an expression of *techne/poiesis* which is a making and building by the human that "has the bursting open belonging to bringing-forth not in itself, but in another (*en alloi*), in the craftsman or artist."[43] In line with the concern for historical epochs, Heidegger is clear that *poiesis* in this sense of "aided bringing forth" has the distinct character of destining, which indicates that in and through the *poietic* activity of bringing-forth, guided by *techne*, the activity instantiates and so reveals the human's historical relationship to the Earth. We note that "working," for the Greeks, "did not involve accomplishing something through strenuous activity, nor was it conceived in terms of cause and effect," for rather the so-called "workness" (*ergon*) of the thing "meant its manner of presencing, *ousia*, *energeia*, the enduring unconcealment of its outward appearance."[44] And it is this form of *poietic-making*, or aided bringing forth, that works in communion with nature and is attuned to the aim of bringing out the best of nature to stand-forth in works of art that preserved and sheltered the original power and mystery of the Earth. The Classical Greeks allowed nature to direct the process of construction, e.g., the sculptor sees and envisions the final form of the figure as already residing in and present to the marble slab awaiting release. Or, the ancient temple architects facilitate the essence of both the stone and wood to shine forth in radiant glory from out of the temple's physical structure. "Createdness of the work," suggests Heidegger, "means: truth's being fixed in place in the figure," but such use "does not use up or misuse the earth as matter, but rather sets it free to be nothing but itself," and it is "at all times a use of the earth in the fixing of truth in the figure."[45] As the ancient Greek temple draws the Earth into the openness of the world, Earth becomes visible. The raw earthen materials of marble and metal acquire a new look as they brilliantly radiate from out of the temple's work-Being, from out of the architectural structure. Marble and metal are transformed within the work, and shine forth in the enigmatic recession and concealment of their presencing. In the work of art (temple) the Earth emerges as a powerful spiritual presence, representing the awe-inspiring sublimity of nature, which rises and exceeds humanity. Heidegger, perhaps most famously, presents this latter instance in his description of the Greek temple in "Origin of the Work of Art."[46] Indeed, the placement of the ancient temples was inspired and dictated by the address of the lay of the land itself, for there was an *originary* sense of responsibility to fit organically into the geographic surroundings instead of disturbing the land. Vincent Scully observes the following as we relate to Heidegger's observations:

> All Greek sacred architecture explores and praises the character of a god or a group of gods in a specific place. That place is itself holy and, before the temple was built upon it, embodied the whole of the deity as a recognized natural force … [T]he formal elements of any Greek sanctuary are, first, the specific sacred landscape on which it is set and, second, the buildings that are placed into it.[47]

The Roman epochal framework

For Heidegger, it was the Romans and *not* the Greeks that established Western civilization. The Roman age is understood in terms of a change to the attunement found in the Archaic and Classical Greek cultures, and this crucially indicates a radically revised view of the human-nature relationship. Whereas the Homeric framework highlighted "wonder" in the face of *physis*, and Classical Greece was expressive of an artistic knowing and making in terms of *poiesis* inspired by *physis*, the ancient Romans approached the building and construction of the Empire in terms of the human stamping out, pressing out, and imposing form on the natural world in a way that prefigures our contemporary understanding and practice of *production* and *manufacturing*. This type of activity is proximally and for the most part unconcerned with bringing out the best of nature through a *poietic* process, but rather demonstrates the instrumental concern with the forging and shaping of things, world, and nature for distinctive human purposes. Here, we get the sense of a superior human subject imposing forms on the external world of nature understood as a collection of objective matter ripe for manipulation. A crucial issue that Heidegger deals with in relation to the Romans is language and the mistranslation and hence disingenuous misunderstanding of many crucial Greek terms. "*Roman thought takes over the Greek words without a corresponding, equally original experience of what they say, without the Greek word.*"[48]

For example, in and through Latin mistranslation, both the human being (*subjectum*) and nature (*natura*) – and as the Latin is taken up by theologians in the Middle Ages, *ousia* becomes *essentia/substantia* – are defined in terms of "things" or "objects" obscuring, and in the extreme losing, the Greek sense of the *coming-to-presence* of phenomena, hence the "interpretation of the thingness of the thing is established," and the "Western interpretation of the Being of beings [is] stabilized."[49] The language henceforth used to speak about both human beings and nature is in a disingenuous manner propositional, which is restrictive and reducible to a "subject" and "predicate," and hence the "structure" of the phenomenon under consideration is envisioned as *only* a "projection of the framework of the sentence,"[50] i.e., known through the correspondence between "idea" and "state-of-affairs" of which the locution expresses. As opposed to saying along with Heidegger that *Being is obscured*, it is possible to state without distortion that such a move obscures the eternal mystery present to all things, most specifically to nature. Even the fourfold structure of the Greeks' understanding of causation is challenged by the Romans – *material, formal, efficient*, and *final cause* – when, according to Heidegger, efficient causality (*causa efficiens*) sets the standard and becomes the sole mode of causation accepted as "that which brings something about," or that which produces a finished product (*ergon*) that "obtains results, effects."[51] The notion of *eidos* or *idea* indicated for the Greeks "'causing' only in the sense of being responsible for bringing something into presence," whereas the Romans understood *aitia* as causation in the sense of "working upon something, effecting it, making it."[52]

Here, as stated, we encounter the essence of *manufacturing*, where the production system is organized around the "subject" setting upon the "object" in order to give it or impose upon it a form so as to serve a decidedly human and *instrumental* purpose. As Zimmerman contends, for Aristotle *energeia* referred to "the self-manifestation or 'being' of an entity," which held ontological significance, but for the Romans *energeia* came to be equated with "cause and effect: the 'action' involved in putting something together or manufacturing it."[53] The Roman manner of building with mass-produced bricks will suffice to illustrate this idea of manufacture in terms of the efficacious *anthropocentric* end of *instrumentality*: The Romans dug white and red clay, that which is formless awaiting a form to be impressed upon it, be that round, rectangular, or triangular in shape, then these bricks were used to give shape and form to their walls, houses, vaults and arches, and baths, to name but five structures.[54] Lost, here, is the original Homeric and Classical Greek relationship between artisan, nature (*physis*), and artifact, which for Heidegger is grounded in the relationship between the human and the essential truth of Being.

The Christian epochal framework

"In the metaphysical God," contends Zimmerman, "Heidegger discerned the form of technological humanity," depicted as "real" or "*actualitas* in the sense of effecting, calculating, planning agency which produces the relatively stable and independent presence of creatures."[55] In short, during the Christian epoch it is God, the divine Creator, who makes all things, and, in the grand scheme introduced in *Genesis*, in unambiguous terms, nature is made for the human being. Here, the "Being of beings" is reinterpreted as "Being-created," in the sense that the human is created in God's divine likeness (*imitatio Dei*) and He then grants them unfettered dominion over the earth/nature and all of its creatures, i.e., "every creeping thing that creepeth upon the earth," and the land, sea, and all manner of sprawling vegetation, "'every herb bearing seed, which is upon the face of all the earth, and every tree."[56] Here, the vast distance between Christianity and the original Greek understanding of *physis*-as-*poiesis* in relation to *techne* and the *technites* becomes evident, for "God's creative work is to be thought of as different from the action of a craftsman."[57] In the Christian epoch, the view is that God Himself, as uncreated Creator, forges *ex-nihilo* the world, humans, and non-human entities: "[B]eings have been created by God – that is, have been thought out rationally in advance"[58] (i.e., *essence precedes existence*), and so the flow or *directionality* highlighting the unfolding of value and *meaning* represents what might be viewed as the *ontological abdication* of the human's *existential* responsibility: God ⇒ human ⇒ nature ⇒ *meaning*. Heidegger argues that as we move into the modern age, the age of secularism and "secular humanism," we reach a point when "the relation of creature to creator is dissolved, while at the same time human reason attains predominance, and even posits itself as absolute, [and so] the Being of beings must become thinkable in the pure thinking of mathematics," and Being as "calculable" in such a manner, "makes beings into something that can be ruled in modern, mathematically structured technology," and this establishes a new (and *essential*) relationship between the human being and the Earth that is grounded specifically in the human's understanding of *tools*, an understanding that is "different from every previously known use of tools."[59]

The modern epochal framework

The modern epoch can roughly be seen as spanning the sixteenth century into the twenty-first century. In this epoch, as stated, the subject becomes deified and nature serves various instrumental purposes via the rise of science and technology. The danger grows, as Heidegger warns, with the death of God and the devaluation of the highest values, when nihilism envelops the modern epoch and the potential exists that the human's unfettered drive to exercise its *will to power* will have cataclysmic effects on the world. A change is effected in the understanding of the human subject, in that humanity becomes the efficient cause of all things objective. With the rise of modern science and technology there is a transformation of "substance from *hypokeimenon* to human subject," for the Greeks saw *hypokeimenon* as that which "lies-forth as the ground which gathers the various aspects of an entity," in terms of the "'basis' or 'substance' of things."[60] However, with the so-called "Cartesian turn," humanity becomes the "substance" of all things, in terms of "the self-certain subject, the grounding ground, the foundation for all truth, reality and value … the efficient cause of everything objective."[61] This is evident, for Heidegger, in the rise of human *machination*, *power*, and *violence*, all of which beget *disaster*.[62] Here, we understand the *attunement* (CF) defining this epoch as the *Metaphysics of Mastery*, which is linked ineluctably with the above analysis of *das Ge-stell* and the "enframing" effect of modern technology. In the modern epoch all objects are used and are in the service of the human subject and its projects. The age of technology is a time of *instrumentality*, *standardization*, and the *mechanization* of *mass manufacturing* and *production*. Let us consider briefly the mechanized contemporary fossil fuel industry as just one example of a systematized means of *technological* production, distribution, and consumption that is directed toward efficacy and economic gain that is attuned to the earth in such a way that nature is transformed into "standing reserve" of "resources" to be exploited for human purposes. For Heidegger, "instrumentality is considered to be the fundamental characteristic of technology."[63]

The contemporary fossil fuel industry, if we stay true to Heidegger, is in the grip of *das Ge-stell*. For in its quest for *mastery* and economic gain, it holds the erroneous and dangerous belief that since fossil fuels are formed over many millions of years in the earth's crust as the remains of dead vegetation are subjected to heat and pressure, such processes are renewable – because they are "natural" – but they are *not*. Because of the extended periods of time required to produce fossil fuels, they cannot serve as an endless "resource" for human consumption: They are in fact non-renewable, and when exhausted such resources are gone forever; their technological essence as "standing reserve" will have been exhausted.[64] Heidegger is adamant that the "setting-upon" of technology "challenges forth the energies of nature [as] an expediting [*Fordern*], and in two ways": As a revealing (a form of *aletheia*), technology unlocks and exposes things; however, the manner in which it accomplishes this "expediting" of things is always itself "directed from the beginning toward furthering something else, i.e., toward driving on to the maximum yield at the minimum expense."[65] The "expense," however, that is to say, the *price paid* or capital invested in technology for securing resources has a crushing effect on both nature and the human being, in that technology's "expediting" ultimately reveals itself, when nature's resources reach the point of near depletion, in terms of the dire ontological *price paid*, the debt and expense incurred, for destroying the planet, e.g., the frightening and irreparable damage done to nature through the processes of extracting, burning, and transporting these fossil fuels. We now turn to the exploration of *secular humanism*, a movement drawing direct inspiration from radical *naturalism*, and to two ecological or environmental movements traceable to it.

Problems with secular humanism and scientism

Critiques of sustainable development and "green" humanism

Secular humanism

Zimmerman provides a chronology of worldviews leading to secular humanism as we understand it, e.g., Francis Bacon's view of science was that it provided the human with unchallenged power over nature. In addition, "Descartes' division of reality into thinking minds and extended matter promoted dualism which stripped the natural world of beauty and intrinsic value," and later Enlightenment views "proclaimed that rationality would make possible a utopian life for humanity,"[66] which included the conquest of nature, portending its future destruction. Bonnett's eco-phenomenology enhances our discussion by labeling the three main characteristics of secular humanism, which he traces to European Enlightenment thought, as follows:

1 It works to elevate, or in the extreme, "deify," the human's faculty of reason to such a degree that it is wrongly and dangerously assumed that nature should be subjugated to human purposes – we expand on this issue in our discussion of the Cartesian and Kantian influences on humanism. Related to this view is the belief that nature is representative of a seemingly inexhaustible store of resources.
2 It harbors a view of nature that is reducible to disciplines and the epistemic structures and categories consistent with mathematics, physics, and biology. When nature is reduced purely to humanly constructed theories and categories, it is wrongly conceived as a phenomenon that has a reality that is imposed on it from the "outside," as it were.
3 As intimately linked to 2: It harbors a mechanistic view of nature as physical matter and energy unfolding because it is given structure according to universal causal laws. This view, as we have already stressed throughout, presupposes that nature is a "mechanical causal or probabilistic system possessing/generating no internal significances and therefore incapable of possessing inherent intrinsic value."[67] This relates to, as introduced above, the mechanistic influence of Cartesian philosophy on secular humanism, which harbors an additional presupposition, namely, that the "structure of reality is to be supplied by impersonal rational cognitive categories and ordering (rather than, say, by emotion, aesthetic sensibility, felt value and worth)."[68]

Introducing Gray's philosophy, we note that Bonnett was the first eco-phenomenologist to engage Gray in the service of education, and in relation to Gray's conclusions regarding *secular humanism*, Bonnett asserts: "Gray is surely right to point out how" the obsession with human progress as a salvific vehicle overlooks the "awareness of mystery and tragedy in human life and a sense of importance of the individual lives (and environments) that are overlooked or sacrificed in the interests of such progress."[69] Within Gray's corpus we encounter a prolific and vociferous attack against secular humanism. However, as a critic of Gray, philosopher Raymond Tallis contends that Gray "despises all technological fixes and any aspiration to improving the lot of mankind."[70] Simon Critchley, in a related criticism of Gray, claims that Gray's philosophy ultimately devolves into the stultifying and unproductive fatalistic philosophy of *passive nihilism*.[71] We do not, however, assimilate Gray's thought *in toto*; we do not adopt all of his tenets, nor do we find them valuable in relation to all of the themes and issues we address. However, we do believe that there is value in Gray's philosophy for inspiring eco-phenomenology, and we locate this in his critique of secular humanism and the destructive relationship with nature it potentially sanctions – this position espoused by Gray is indeed consistent with the critical views of radical *naturalism* in the eco-phenomenological movement, although Gray's philosophy is referenced, as stated above, only by Bonnett. It must be noted that we part ways with Gray concerning the belief in the transcendent potential of the human being for its continued development, a notion Gray eschews, and we reject the radically skeptical view that categorically distrusts and hence discounts the value and importance of the sciences. Our position is that we need to, *because we can*, *by freely choosing to do so*, reconceive and transform our relationship to, not only nature, but also to the science and technology employed in our contemporary dealings with nature.

To begin, we must be cautious when using the term "secular" in relation to Gray's critique of humanism, which for Gray, is synonymous with the erroneous *belief in human progress as mechanism for salvation*, i.e., "by using the new powers given us by growing scientific knowledge, humans can free themselves from the limits that frame the lives of other animals."[72] There are a variety of dangers lurking in this observation, not the least of which is the erroneous and misplaced belief that human knowledge will continue to develop, and so the human's power will subsequently increase. However, according to Gray, this is not so, as the radically ego-centric and anthropocentric drives of the human (*homo rapiens*) will not change, and so the human's predisposition for behaving in ways that are utterly "destructive and predatory" will be empowered and elevated to new levels of efficiency in the control and mastery of nature in its quest to feed its insatiable appetite at all costs. Gray importantly brings our attention to the role of positivist thought, and although "Positivism is largely forgotten today, it was more influential than any other intellectual movement in shaping the humanist creed" that succeeded Christianity as the new "ready-made worldview."[73] The cult of positivism provides liberal humanism with beliefs that are leftover remnants from a faltering Christianity, where, as opposed to the Judean-Christian God, as already stated, it is the human that is deified, and hence it is conceived as different than and superior to every other animal. Humans declare that they are unique in that they hold the untenable belief that by "using the power over nature given … by science, [they] can create a better world than any that has existed before," and in this view, a running and indelible theme, as it relates directly to our eco-phenomenological concerns, "the earth is simply a mass of resources for human use, and the other animals with which we share it have no value in themselves."[74] Gray is critical of the manner in which this view influenced both the interpretations and the political enactments of Karl Marx's philosophy, which inspired "policies that resulted in environmental catastrophe in the former Soviet Union," e.g., the "destruction of peasant farming and its replacement by agriculture collectives," which led to the "dust bowls and famines of the 1930s."[75] According to Gray, to provide but one example, it was the misreading of Marx along with the influence of positivism that spawned the attempt to heedlessly master nature, which as Gray argues, contributed to the eventual downfall of the Soviet Union.

Unique to Gray's argument, as we address below, is the revelation of the Christian (religious) eschatological roots that give structure to what we have referred to as "secular" humanism. The periodical *Free Inquiry* (published by the Council for Secular Humanism) advances 21 tenets or "affirmations of humanism," and states in unequivocal terms that humanists are committed to battling any "efforts to denigrate human intelligence" while resisting the tendency to "explain the world in supernatural [religious] terms." Indeed, Jean-Paul Sartre's existentialism is a decidedly "atheist" or *secular* form of humanism,[76] and when discussing what might be termed the three basic "moods" of existential philosophy, Sartre talks of the manner in which the human, as *Being-for-itself*, shoulders the responsibility for living a life of *anguish*, *forlornness*, and *despair*. As Sartre asserts, *forlornness* indicates that the human being is estranged from objective values – or the *luminous realm of values* – and must accept the ontological certainty that "God does not exist and that [humans] have to face all consequences of this."[77] In addition, when Sartre elaborates his meaning of *despair*, he claims that humans should confine themselves "to reckoning only with what depends on [their] will, or on the ensemble of probabilities which make … action possible"; indeed, "no God, no scheme, can adapt the world and its possibilities to [the human] will."[78] To ignore this and, for instance, to wish or hope for divine or supernatural intervention during troubled times, and hence depend on something other than one's own will and one's own unique possibilities, is to live in "bad faith." As should be obvious from these brief examples, the moniker "secular humanism" as described opposes any and all religious, or what we might identify as supernatural or mythological, explanations of the world.

Above, we spoke of the influence of positivism on inspiring the type of *scientism* that underlies "secular" humanism, but Gray reveals the fact that humanism also draws a great deal of its strength, lasting-power, and appeal from the Christian faith, with its drive toward utopian and eschatological beliefs, or *end of history* (apocryphal-"end-times") thinking. Humanism, argues Gray, is a contemporary "secular mutation of Christian apocalyptic beliefs,"[79] in which the Christian myth is transformed and rendered "into the idiom of science."[80] However, according to Gray, humanism is far more irrational than the Christian belief system and, as related to our concerns with nature, it is also far more dangerous and harmful to the planet. As related directly to our brief analysis of Gray's criticism of humanism, it benefits our position to recall Albert Camus's sustained critique against modern manifestations of failed utopian *revolutions* and metaphysical *rebellions*, all of which harbor in a tacit and repressed manner, Christian ideals: Bourgeois formalism, Marxist revolution, and fascism or Nazism. Each of these, as David Sprintzen reasons, had its own unique way of attempting to represent an "instrument of human salvation," e.g., in Marxism we find it in the deification of the "'rational' product of a scientific analysis of the historical process that revealed the mission of a class and the inevitability of its ascension."[81] Here, in a similar manner to our description of secular humanism, Camus recognizes that these movements were "deifying material progress, working through an open-ended linear history," and thus that they all "remain[ed] wedded to transcendent Christian values, reduced to a pure formalism," and ultimately, much like liberal or secular humanism in Gray, obscured or "cover[ed] an unrepentant materialism and social exploitation."[82] The Christian views God as bringing about a "happy end" to history, whereas various outgrowths and manifestations of humanism "promise something more miraculous,"[83] namely, that by harnessing and employing the power of science and technology, human beings hold the potential to bring about an "eschatological" historical time when humanity's problems are solved through the establishment of a *technological* utopia on the earth; and humanists hold the inflated and hubristic belief that science will allow them to accomplish this end more efficiently than any god. Science, claims Gray, in relation specifically to its manifestation in *scientism's* belief in the rise and dominance of human reason, is not "producing a more rational view of the world," but in reality is producing and perpetuating "a secular mythology that is further from the truth of the human condition than the religious myths of the past."[84]

Naturalism-as-scientism

As we have already stated, *scientism* is not a discipline associated with the natural sciences; rather it is the belief – forming an overarching belief-system or *Weltanschauung* – in the unadulterated power of science and empirical methods to explain all that is knowable about reality, or as Raymond Tallis makes clear, it is the "mistaken belief that the natural sciences can give or will give a complete description and even explanation of everything, including human life."[85] We experience the rise and dominance of *scientism*, as Detmer argues, in "recent attempts to reduce truth claims in various object domains to explanations rooted in some scientific theory,"[86] e.g., biology, sociology, economics, or neurology and the cognitive sciences. Indeed, in addition to Husserl, Heidegger is aware and critical of this view when he dismisses the suggestion that the possibility exists to construct a complete "scientific" image of the human being "in knowledge" by bringing together the findings of various academic disciplines, such as the natural sciences, social science, psychology, anthropology, and the like. According to Bonnett, scientism is *epistemologically exclusionary*, because it dogmatically embraces the unwavering view that scientific thinking "should be the arbiter of [all] good thinking as a whole and that it holds a privileged position when it comes to identifying and investigating fundamental questions concerning the nature of reality,"[87] which as related to Gray's philosophy, constitute the means that are necessary to secure *human progress*. Drawing from our analyses of philosophers, phenomenologists, and eco-phenomenologists, we understand that the predominant CF at work providing form, meaning, and the en-framing *gestalt* for the structuring of our relationship with the world and nature is grounded in the view to *scientism* or *radical naturalism*, which is expressed through a pernicious form of "humanism" that is anthropocentric; and this view, as we have seen, is expressed in the unchecked hubristic belief in unlimited human progress (indeed, unimpeded human transcendence) based in the "faith" of the continued advancement of science and technology. According to the critical view of eco-phenomenology, in the grip of a humanism that is dominated by *naturalism*, our understanding of human rationality becomes reducible to and inseparable from the "scientific method," which wrongly indicates that rationality, as an expression of a "hard" scientific *naturalism*, holds the power to describe, explain, and predict outcomes with ruthless efficiency, in a way that reinforces a truth in the categorical certainty of the claims that emerge from this form of rational thinking.

According to Brown, the form of *naturalism* represented in *scientism* falls victim to "dualistic thinking or Cartesian ontological categories," but in a sinister twist, while accepting the distinction between *res cogitans* and *res extensa*, it denies the "reality of mind, spirit, value, and meaning – leaving only a world of value-free physical entities, namely, *res extensa*,"[88] which Brown identifies and labels *Cartesian monism*. In this view, which is antithetic to the position adopted by eco-phenomenology, material entities are considered real and thoughts concerned with value and meaning are thought to be unreal and wholly subjective in nature, and hence untrustworthy and expendable. This model holds crucial implications for our move to the realms of *aesthetics* and *ethics* in our eco-phenomenological exploration of nature as we push forward into Chapters 3 and 4. The danger exists, as Lester Embree observes, that when we are focused on nature as objective, "values and uses might then be considered not to matter at all or not even to exist," as they are "considered artificial determinations added to naturalistic objects."[89] This view of "objectivity" fostered by *scientism*, providing structure to the modern CF, has been further shaped "by Descartes' interpretation of [phenomenon] in terms consistent with the objects of a mathematized Galilean science,"[90] and so includes the following erroneous assumptions: (1) The world admits of a single, overarching, and correct explanation, (2) The world is comprised of determinable and identifiable properties, dancing to the indelible laws of antecedent cause-and-effect, and (3) The world lives as an objective, external, mind-independent phenomenon that can be understood – as in (1) – with a high degree of rational certainty. As introduced earlier, the phenomenological critique of *scientism* seeks a return to the "life world," which would reveal, as Embree stresses, "the metaphysical privileging of *res extensa*," in a way that would bring our attention to the error we fall victim to when mistaking "what is a complex multileveled historical abstraction for the nature of reality."[91]

In relation to this specific issue, as we have thus far analyzed, all expressions of eco-phenomenology share the view that what we understand as the CF is given structure by attitudes and understandings consistent with a Cartesian worldview. To reiterate, when reason is privileged as the superior human capacity with which to know the world, and when nature is understood in terms of a causal-mechanistic "machine" – nature as the *objective* and *external other* – it reinforces the catastrophic anthropocentric drive (the technical *will to power*) to master and subjugate nature in an *instrumental* manner. Importantly, we argue that *scientism's* expression through the CF might be more accurately understood when associated with, in addition to Descartes, Kant's philosophy in the *Critique of Pure Reason*. Recall that for Descartes, it is God and *not human reason* that ultimately *brings the world*. Simon Critchley contends that Kant's philosophy marks the schism dividing analytic and Continental philosophical thought, and this division occurs, and is still acknowledged, depending on how Kant is read, e.g., is the reading limited to Kant's First Critique or the *Critique of Pure Reason* or is it inspired "by the greater systematic ambitions of the Third Critique or *Critique of the Power of Judgment*"?[92] On Critchley's account, the first reading is focused primarily on epistemology and the transcendental subject's relation to the world in and through knowledge, and here the concern is with whether or not Kant "successfully provides a valid foundation or grounding for empirical knowledge that meets the challenge of the skepticism of David Hume."[93] In this way, Kant is understood as making a contribution to the philosophy of science, and indeed this is how Kant is read by a majority of analytics or Anglo-American philosophers. However, if Kant is read in light of the Third Critique, "then the burning issue of Kant's philosophy becomes the plausibility of the relation between pure and practical reason, nature and freedom, or the unity of theory and practice," and these issues are linked to ethics, aesthetics, and human autonomy. This is "precisely the route followed by the German idealists," e.g., Fichte, Schelling, and Hegel, and it is "arguably the route that Continental philosophy has followed ever since."[94]

Let us briefly explore how Kant's First Critique might be understood as establishing the relationship between the human and the world of nature in terms of the *epistemological directionality*, instantiated in what we shall call the "cognitive correlation" between human thought-and-world, where the objective world conforms to human concepts and ideas and not the reverse. Thus, the movement or *directionality* of our understanding is represented in "subjectivist" terms (Human⇒world⇒*meaning*). In Kant, systematic a priori judgments are possible by means of pure intuitions, which are linked with a priori *synthetic judgments*, e.g., time, space, and causality are not located "out there" in the external world; rather they are in the mind in terms of "categories of the mind," or understanding, and they provide structure and form to the perceptions we receive from the external world. Since all humans possess these so-called "categories of the mind" – we are all similarly *hard-wired*, as it were – these give rise to Kant's notion of the *transcendental subject*, and further, because the mind produces ideas of the external world, and the mind and its ideas give form to that world, Kant's philosophy, a bridging of the gap between rationalism and empiricism, is classified as *transcendental idealism*. Kant states this position, regarding the *epistemological directionality* of knowledge as described above in relation to the world, in the following passage:

> Hitherto it has been assumed that all our knowledge must conform to objects. But all attempts to extend our knowledge of objects by establishing something in regard to them a priori, by means of concepts, have, on this assumption, ended in failure. We must therefore make trial whether we may not have more success in tasks of metaphysics, if we suppose that objects must conform to our knowledge.[95]

Whereas Copernicus places the sun at the center of the universe by de-centering and displacing the earth, as eco-phenomenologist David Jardine observes, Kant establishes human reason as the "center of the *knowable* Universe," and reason determines what is knowable, for reason as a "synthesizing faculty, bestows order upon the universe, turning the chaotic influx of experience into a knowable (i.e., orderly) *cosmos*" – reason in this sense is "*constructive of a world*."[96] This view not only has the potential for framing a mindset that determines our experience of nature, giving rise to behaviors that exact catastrophic ecological outcomes, it is our contention, based on our analysis of the CF, that this view is already at work attuning us into these reductive modes of thinking and destructive patterns of behavior. For Kant's philosophy gives us a view of the subject that is estranged and completely separated off from nature, and this subject no longer dwells within nature in terms of an *immersed*-intentional denizen, with the potential to assume the role of nature's guardian, but rather stands against it as subject before an object. According to Jardine, Kant "embodies the denial that [nature] somehow guides us, sustains us, and bears us up" and makes us the *children of nature* – in Kant, as Jardine solemnly observes, "ecological disaster is already foretold."[97] For in and through the Kantian legacy "the spirit of [nature], its inner fire, its *logos*, becomes ourselves. Eco-logos has become ego-logo."[98] In and through Kant, nature must present itself to the human and it is human reason that "creates and controls and delimits the 'space' within which things are brought into their own and flourish," and since nature "gains integrity from us, we can colonize it without hesitation, for this simply fulfills what it already prescribed *a priori*."[99]

Sustainable development, green humanism, and scarcity

It is crucial to keep in mind, as we have already shown in our analysis of Deep Ecology from Chapter 1, that even well-meaning philosophical views that inspire deep "ecological" movements seeking to "preserve" nature or the planet tend to operate out of the en-framing that is antithetical to the relationship with nature that eco-phenomenology embraces, for such views as Deep Ecology still draw their life-force from, as related to our view of the CF, a "highly pervasive metaphysics of mastery," and the concomitant "arrogant instrumentality that this entails hollows out our relationship to what is present."[100] According to Bonnett, even what he terms "bio-centric views" harbor a latent anthropocentrism, and while they "may be suggestive of important alternative attitudes ... they lack the intellectual resources to replace it."[101] The problem with such views (sustainable development) as we discuss next, is that they are focused on what amount to superficial changes to the human's view of its interaction with the environment, amounting primarily to the concern for preservation in order to promote and prolong our own thriving on the planet, while remaining unconcerned with more radical and substantial, and shall we say, deep "philosophical," changes to our attitudes or *ethical dispositions* which, as consistent with the eco-phenomenological project, require the re-establishment of the "legitimacy of modes of thinking and being that the metaphysics of mastery occludes."[102] For example, the ecological philosophy underlining "sustainable development" embraces the Utilitarian principle of posterity (*problem of posterity*), which holds the view that it is possible to develop the earth in such a way that the planet's resources meet both our present needs as well as the needs of future generations. Thus, our use of resources is accomplished in a way that avoids impeding future generations from also meeting their needs. However, as Kerry Turner recognizes, sustainable development can never adequately or realistically meet the needs of humans in a modern industrialized economy for, as stated earlier in relation to fossil fuels, certain resources are finite and exhaustible and cannot be renewed: "It makes no sense to talk about the sustainable use of a non-renewable resource," argues Turner, for even with recycling and the strict regulation of the extraction, consumption, and distribution of resources, "any positive rate of exploitation will eventually lead to exhaustion of the finite stock."[103] Gray echoes this sentiment, for as population growth continues to stress the environment and its limited supply of resources, humans are "producing environmental changes on a scale unprecedented in history," and as related to problems that we have associated with sustainable development, with respect to dwindling resources, Gray makes the shocking and disturbing observation that "in parts of the world [even] fresh water is becoming a non-renewable resource."[104]

Relating to these foregoing remarks, Bonnett observes that, on the surface, sustainable development appears to hold positive potential for a renewed relationship to the planet and the use, distribution, and preservation of its precious resources; however, as he punctuates, this amounts to a generous and highly charitable reading, for sustainable development appears to weld together "two highly desired goals": (1) The notion and practice of consumption-and-conservation, which seeks to address the seemingly intractable issue and problem of valuing Nature while at the same time serving continually expanding human demands and needs; and (2) The "idea of accommodating ongoing human aspirations to 'develop' that is, in some sense to have more or better."[105] Although this view appeals to both "enlightened captains of industry" and "eco-warriors," there is an inherent problem with it: When questions and concerns are raised regarding *what is sustained, at what level, and over what time span*, and in addition, regarding *whose needs are being met, how they are prioritized, and according to what criteria*, the answers and responses, when provided, are questionable at best and disturbing in the extreme. This is because sustainable development embraces, as is consistent with *secular humanism*, anthropocentric, instrumental, and economic principles that are still working to reduce nature to a vast resource store for human consumption, "an object to be intellectually possessed and physically manipulated and exploited in whatever ways are perceived to suit (someone's version of) human needs and wants."[106] Thus, sustainable development might be understood as yet another version or modern manifestation – albeit a "soft" version, so to speak – of the technical-scientific *will to power*, the goal of which is the mastering of nature in the present as well as the future with the explicit purpose of sustaining the human race and insuring its progress. Since the aims of sustainable development are instrumental and anthropocentric, primarily grounded in a scientific view of the world (that is expressed as a veiled form of *scientism*), what is missing or occluded in this view, because it is primarily concerned with knowledge as the means by which to master and exploit the world for human resources, is a grounding ontological fundament. In addition, although couched in terms of utilitarian ethics, sustainable development smuggles in a scientific epistemological view that severely limits the scope of what counts as "valid" or even legitimate knowledge.

What Gray labels "Green Humanism" is identified as yet another manifestation of neo-liberal humanism, for although they know that the earth can never be mastered, within their "Luddite struggle against technology they renew the illusion that the world can be made the instrument of human purposes."[107] The utopian ideal that emerges from Green Humanism is the myth that salvation can be found in humans taking charge of their destiny by adopting an attitude of guardianship, but this form of so-called "guardianship" is grounded in the capitalization of nature's resources. However, just as in sustainable development, there is a latent anthropocentrism lurking at its core, when the reality is that any legitimate potential to save the planet lies beyond the severely limited powers of the human being. Gray observes that humans "struggle to conserve what is left of the environment … moved by the love of living things, *biofilia*, the frail bond of feeling that ties humankind to the earth."[108] However, this is an illusion, because "the mass of mankind is ruled not by its intermittent moral sensations," but more realistically by the fleeting "needs of the moment"; and in light of the reality of the scarcity of resources, as alluded to earlier, the human "seems fated to wreck the balance of life on Earth – and thereby to be the agent of its own destruction."[109] In the extreme, a view with which we strongly disagree – as we have seen in relation to Naess's Deep Ecology – Gray claims that the only thing that will truly rescue the planet is extinction of the human race. "Humans will never initiate a relationship of mutually beneficial symbiosis with [nature]," for as Gray pessimistically concludes, the "advance of *homo rapiens* has always gone with the destruction of other species and ecological devastation."[110]

To conclude this section, let us return to Brown and Toadvine, who urgently stress, "environmental destruction and crisis are caused by core beliefs within our worldview," which we have referred to as the CF, "that sanction, legitimate, and even encourage the domination and technological control of nature."[111] As is evident, our eco-phenomenological project works toward a renewed and reconceived relationship between humans and nature, and while we disagree that *human extinction is the solution to saving nature*, we wholeheartedly agree with Gray's concerns related to the all-too-real problems traceable to the scarcity of resources. "The earth's resources are finite; they cannot be stretched to accommodate the unlimited wants of an exponentially growing population," and human progress will never be able to "transgress the limits imposed by natural scarcity."[112] The belief that resource scarcity can be overcome through industrialization or new forms of technology is an erroneous and contradictory view that is common to global politics. Scarcity is perhaps the most pressing issue we and future generations face, according to Gray, an issue upon which the future wars of the world will most certainly be waged in the Middle East,

China, the US, and Great Britain, as the people of these regions desperately and arduously strive to acquire and accumulate the limited wealth of the planet for their own ends. As we witness today with monotonous regularity, this ignites, to cite but one example, the fierce competition for the waning reserves of oil and natural gas. The "intensifying competition for natural resources flows inexorably from worldwide industrialization – a process that is seen by practically everyone today as enabling humanity to emancipate itself from scarcity," when it is in fact the case that industrialization itself is central to perpetuating and accelerating the problem, which unfolds in the manner of a *vicious circle*, it "creates a new set of scarcities,"[113] it is the *problem* masquerading as the *solution*.

The ongoing influence of the "*naturalistic*" cultural framework

Fossil fuels and environmental politics

Recall our earlier discussion of the *natural attitude* as one mode of disclosure comprising the overarching "life world," recall also our analysis of the three dimensions of Husserl's phenomenology – if the dominant CF is attuned primarily by a radical *naturalistic* worldview, then *naturalism* attunes the human's modes of comportment to its world. The reasons why we do things, along with the concerns and values we hold and instantiate, influence our attempt to interpret and give structure to our world as we comport ourselves in activities that have the very *real-world* potential to change the course of the planet. Thus, if attuned primarily by the *natural attitude*, which has been dominated by *naturalism/scientism*, comportment is enacted – and a concomitant relationship to nature is established – as it is determined by a mindset emerging from out of, because it is intrinsic, to the *metaphysics of mastery*. Conversely, the attunement of the "life world" also holds the potential to contribute to, as it informs the unfolding of, a form of comportment that is attuned and receptive to that which swells or rises up from out of the depths of a sublime experience of nature, an experience that is more in line with and expressive of an eco-phenomenological view of *immersed*-intentional comportment (*corporeal intentionality*). In this view, as we demonstrate, nature speaks to us, and if our mindset has been transformed we anticipate and are receptive to its call or address in ways that inspire rejoinders that are ethical in nature, responses that are in harmony with the tone, timbre, and pitch of its primordial, resonating voice. The chapter concludes with a reading and analysis of the influence of the CF on the "life world," grounded in the *voluntaristic* politics at work in the US that is emblematic of the type of *radical naturalism* of which we have been critical, by exploring the fossil fuel industry and contemporary environmental politics, which we claim are motivated by drives that are *instrumental*, *economic*, and *anthropocentric* in nature, expressing an attitude that is neither *protective* of nor *respectful* to the environment, to nature.

When exploring the role of politics in phenomenology and eco-phenomenology, we are focused on the supreme and predominant role that human autonomy plays in this realm of communal involvement, and what we are concerned with, which can be drawn from later Husserl, is the phenomenon of *metaphysical voluntarism* at work giving structure to the "life world," or as we have introduced, the CF. As Michael Gillespie points out, in early Husserl there was little concern with society or the collective life of a people, and at most any social or political concerns, when present at all, were either indirect or secondary. Husserl originally saw the *natural attitude* as something that tacitly communicated our political and communal views of the world, but later came to see that to establish "philosophy as a rigorous science depends upon a prior understanding of the intentional structure of communal life."[114] This indicates that the way in which we comport ourselves collectively – in terms of how we are attuned – shapes and "modifies the practical/political world that we inhabit."[115] And, as related to our notion of the CF's global influence, Lester Embree stresses that "environmental politics can be practiced on international, national, and regional levels,"[116] and the CF thus acts to form and transform collective social attitudes of those inhabiting the *Lebenswelt*. In Brown's eco-phenomenology, in the discussion on environmental philosophy, we learn that

> both Radical Ecologists and Husserl, argue that the very idea of nature is deeply historical, in that the founding metaphors we use to construct a concept of nature are neither universal nor necessary but rather are historical achievements of particular cultural life forms.[117]

So, in relation to what was introduced in Chapter 1 and referenced throughout this chapter, the phenomenological project, as in eco-phenomenology, is focused on attempting to dig beneath the surface layers of historical interpretations in order to reveal a more original way of Being-in-the-world. Eco-phenomenologists such as Gillespie argue that contemporary politics, and here we should recall what Gallagher earlier claimed about positivistic science and philosophy, acts in one way or another as an "impediment to philosophical activity."[118] By searching out the phenomenological realm of *immersed-intentional* dwelling, we might "recapture the immediacy of existence, the intensity of living that is lost in the complicated mediating institutions of political life,"[119] and although this might not result in a lasting or stable new sense of the political, it might serve to contribute to, recalling Husserl's critical philosophical project as related to the modern sciences, "the continuing voice of dissatisfaction with the politics of everyday incarnations."[120] In line with our account, Embree also argues that valuing, or what we understand in terms of a *sense of normativity*, does not predominate our current involvements with the environment or Nature, and that rather there is an acute sense of the *volitional* that primarily organizes and inspires our interactions with the world and others.

Contributing to our analysis of the influence of the CF, the two types of willing Embree links with environmental destruction or exploitation are wholly consistent with our understanding of secular humanism, namely, *willing for* and *willing for the sake of something else*: In each of these views "the object encountered volitionally can have the characteristics of being positive or negative ends or means."[121] The expression of willing in the practical sphere (*praxis*) primarily assumes the form of human *machination* as expressed through "technology," which requires the use of machines and tools to influence the course of events and change the shape and direction of our existence, e.g., as we currently encounter in the environmental and genetic engineering of plants and other food sources. In the process of humans willing the world there is always an encounter with *objects-as-volitionally-encountered*, and it matters a great deal how this encounter is structured and for what purposes objects are confronted and volitionally engaged and manipulated. The question regarding what are we *willing for* and *willing against* becomes a crucial issue, and so we must adopt a critical attitude when analyzing the valuing system that underlies our volitional choices. As is evident from our foregoing analysis, the willing that underlies modern instances of *voluntarism* is for the most part grounded in what we have identified as the Cartesian-Kantian underpinnings giving shape and life to the modern CF, which indicates that willing remains radically anthropocentric and, in such a valuing system that inspires both our *willing for* and *willing for the sake of something else*, the environment presences in terms of something to be exploited for human purposes. Let us move to briefly explore the attunement, sway, and influence of the CF as we have presented it for analysis, and to explore the idea of *volition* and its origins and effects. In doing so we highlight and attempt to concretize the evidential and ongoing effects of the CF on the fossil fuel industry, which often includes devastating accidents associated with the transportation of fuel and crude oil, as occurred in Prince William Sound in the US as a result of the now-infamous Exxon Valdez disaster or oil spill of 1989.

The fossil fuel industry is inseparable from Capitalism, and Amy Leather traces the origins and rise of the industry back to Great Britain's transition to fossil fuels in the early years of the 1800s, a move that ultimately created the dependence on, for our purposes, oil and with it the rise to supremacy of the big oil companies, representing a dominant economic force that continues in the contemporary global economy. The thought of a transition from fossil fuels such as coal, gas, and oil to more clean and sustainable forms of energy such as solar, wind, and hydro is an astronomically costly endeavor, and it is impossible to imagine, and the testimony of history substantiates this, that companies operating in service of the oil industry "would be happy to write off that amount and replace it with a renewable energy system."[122] There is a governmental and deep-seated historical investment in fossil fuels that engenders the power to lobby governments and their elected officials to sway favors in matters of policy, which more often than not, in a disturbing historical trend, results in the establishment of policies that seek to increase profits for large corporations, but that are highly destructive to the environment. In short, at the time of this writing, politicians in the US are accruing large sums of money to push the agendas of Big Oil in order to get elected and then in turn pledge their loyalty and offer their service, in a move of acquiescence, to the demands and interests of the oil companies, which in many cases are corrupt entities. Indeed, the widespread political corruption in Washington, DC as related to the fossil fuel industry is well known and well documented; in addition to proposing policy it involves repealing laws that ultimately cut into the profits of greedy corporations and manufacturers who are legally forced to cede to Environmental Protection regulations put in place to ensure that the environment and the human population are made safe from the ill, and even fatal, effects linked with the extraction and burning of fossil fuels.

To understand this foregoing claim, we need only to look at the current Trump administration in the US, which champions and enacts this worldview (CF) through its heedless policy-making grounded in the dangerously misguided view, steeped in the venom of licentious avarice, that the Environmental Protection Agency (EPA) stands in the way of the unimpeded economic growth of corporations and big business, and that capital gains of the ruling class (the top 1 percent) outweigh the very real and tangible destruction done to both the environment and the populace of the country. Indeed, on June 1, 2017, President Trump withdrew the US from the 2015 Paris Agreement on Climate Change mitigation, making the following outlandish claim: "The Paris accord will undermine (the US) economy," and "puts (the US) at a permanent disadvantage."[123] In response to the monumental ecological crisis we face, to highlight another issue, despite the fact that global warming is a scientifically established phenomenon, Scott Pruitt (former disgraced, scandal-plagued Chairman of the EPA 2017–2018) is a staunch climate change denier who espouses a radically anti-scientific worldview. Unfortunately, we recognize that Pruitt's replacement, Andrew Wheeler, is a former fossil fuel lobbyist "who shares Mr. Pruitt's zeal to undo environmental regulation."[124] The overarching goal of the administration is to, as efficiently as possible, with reckless abandonment, roll back and repeal as many sanctions and environmental regulatory policies as possible that hold individuals and corporations responsible for complying with the laws that continue the necessary efforts to keep the air, water, and land (nature) clean, healthy, and habitable for humans, fauna, and fowl.[125] In short, their goal is to repeal as many environmental protection policies focused on ensuring clean air, water, and land that were drafted and passed as law by the outgoing administration of former US president Barack Obama.

As related to the subject of fossil fuel, specifically oil, what follows are three policy decisions among many by the Trump administration that put the environment and humans at greater risk, escalating damage to the landscape, natural habitats of non-human species, the water supplies of local denizens, and the quality of the air citizens of the US are breathing: On February 7, 2017 the Trump administration approved the Dakota Pipeline; on March 24, 2017 the Trump administration approved the Keystone XL Pipeline; and on April 5, 2017 the Trump administration withdrew guidance for federal agencies to consider greenhouse gas emissions in National Environmental Policy Act Reviews. Amy Leather concludes that in the twenty-first century, which is now experiencing the heightened deleterious effects of the continual and widening extraction and burning of fossil fuels for over two centuries since the early 1800s, we have seen global energy consumption tripling post-1945, and "the impact of climate change intensif[ying] all struggles bred by capitalism," and in this time of exigency, we not only have to fight "the further expansion of fossil fuels, but we also need to fight to replace a society based on accumulation for profit with one based on production for need."[126] Here, we undoubtedly encounter, albeit in a tacit manner, the support of *sustainable development* in the socialist philosophy of Leather, and recall, based on our earlier critique, the problems inherent in this view as related to eco-phenomenological concerns. However, based on what she and others are describing and documenting in their articles on the devastating influence of fossil fuels on the planet, it is possible to envision the modern CF – as *metaphysics of mastery* – shaping, guiding, and driving these historical processes.

There has been a shift over the last few decades in the fossil fuel industry from underground mining to surface mining, which often employs strip mining and mountain top removal – see Chapter 5. Surface mining strips off the upper layers of the earth to access coal, and in the process it does great violence to the areas that are mined, e.g., stripping trees and vegetation from hills and then using explosives to blast away hundreds of feet of the earth. In the immediate aftermath, massive accumulations of rocks and soil are dumped into valleys and streams in the area, diverting the flow of rivers and streams, radically altering ecosystems. In the long term, sites previously used for strip mining are left with poor soil that does not support the regrowth and repopulation of the once indigenous plants that were destroyed or removed during the process. "The EPA reports that as of 2010, mountaintop removal coal extraction had buried nearly 2,000 miles of Appalachian headwater streams, some of the most biologically diverse streams in the country."[127] Surface mining also affects the developmental health of the greater surrounding eco-communities, e.g., flashfloods, mudslides, and landslides are common occurrences. When the toxic chemicals produced during the mining processes, such as selenium, arsenic, manganese, lead, iron, and hydrogen sulfide seep into local waterways, they not only pollute the drinking water supply, but also harm all the animals, aquatic and land varieties, that depend on the waterways for their existence.

There are also grave environmental health risks associated with drilling for oil and gas both inland and offshore, e.g., the practice of "fracking" for natural gas using hydraulic fracturing requires potentially dangerous chemicals that more often than not remain undisclosed to federal regulators. When oil and gas are extracted, the water trapped in the geological formation comes to the surface, carrying with it dissolved solids, hydrocarbons, and radioactive material that make it difficult to dispose of safely and also unsafe for human consumption as it seeps into the water table and finds its way into the drinking water supply. It is also essential to note that drilling wells, building access roads, constructing processing facilities, and building and installing pipes to transport the oil also disturbs huge swaths of land. In addition to all the fossil fuel that is burned by consumers and industries, in private and commercial sectors, the process of extracting the fuel also produces gases that are harmful to the ozone layer, such as methane and carbon dioxide, and it is crucial to note that methane is a far more deadly greenhouse gas than carbon dioxide. Besides air and water pollution, as Melissa Horton notes, "fracking also increases the potential for oil spills, which can harm the soil and surrounding vegetation," and fracking can also potentially "cause earthquakes due to the high pressure used to extract oil and gas from rock and the storage of excess wastewater on site."[128] The transportation of fossil fuels holds many potential dangers, for when "oil is transported across the ocean in supertankers, and it is moved over land by pipeline, rail, and truck," there exists the ever-present risk to the environment that oil spills pose, and it is the case that oil spills from tankers and offshore oil rigs pose some of the greatest and most imminent threats to our oceans and ecosystems,[129] as is evident by the many crude oil spills that have occurred in recent history.

To this very issue, on March 24, 1989 the oil tanker Exxon Valdez struck the Bligh Reef in Alaska spilling over 11 million gallons (262,000 barrels) of crude oil into Prince William Sound. Until the 2011 Deep Water Horizon spill in the Gulf of Mexico, it was the largest oil spill in US history. Megan Gannon reports on the continued and sustained effects of this disaster, and it is shocking and horrifying that even now, close to 30 years since the spill, oil can still be found trapped beneath rocks – in the lower intertidal zone: "Heavy residual oil is still trapped in sediments of Smith Island."[130] To this present day the area has not fully recovered, and this oil continues to eradicate mussels, clams, and other marine life, leading to thinning populations of animals dependent on these species for food, hence greatly disrupting the local ecosystem as a result of long term devastating effects. In all, as David Williamson reports, more than 1,000 miles of coastline were affected and hundreds of thousands of animal species perished, e.g., an estimated 250,000 seabirds, 3,000 sea otters, 300 harbor seals, 250 bald eagles, and 22 killer whales died immediately after the spill occurred, "species as diverse as sea turtles, harlequin ducks and killer whales suffered large, long term losses," and in terms that are staggeringly horrific, related to and amplifying what was stated at the end of the previous paragraph, "the oiled mussel beds and other tidal shoreline habitats will take an estimated [additional] 30 years to recover."[131]

Corporate greed manifested itself on many fronts in ways that inevitably contributed to the disaster, and there was a startling lack of concern by the Exxon Corporation for the magnitude of the devastation the oil spill caused to the Alaskan shoreline when the oil poured from the gaping hole torn in the tanker. Daniel Read reveals that highly inexperienced and understaffed crews on the tanker were enduring grueling working hours, e.g., "the Exxon Valdez had just 19 crewmen to transport its 53 million gallons of oil in a vessel almost a thousand feet long."[132] Indeed, the ship's captain was not even at the helm when the tanker rammed the reef but was later discovered in an inebriated state in his cabin; this left an inexperienced seaman – the third mate – to command the tanker. The ship's Collision Avoidance System radar (RAYCAS) was not functional and so failed to alert the third mate to the potential collision. As Greg Palast reveals, "the tanker's radar was left broken and disabled for more than a year before the disaster, and Exxon management knew it … it was just too expensive to fix and operate."[133]

Prior to the disaster, Exxon focused on dramatically increasing its profits by regularly cutting corners in a variety of ways, e.g., the oil company flagrantly ignored federal regulations for the installation of state-of-the-art iceberg monitoring systems on the ship in an effort to keep its expenditure to a minimum while ensuring that profits soared. Exxon was "routinely running understaffed and poorly equipped tankers," and the corporation's "disaster response teams were also inadequately outfitted, which played a part in the spill turning into such an unmitigated catastrophe."[134] In all, safety, efficacy of transportation, and concern for the environment were supplanted with the overarching concern for increasing Exxon's profit margins. Read also informs us that Exxon was slow to act in terms of cleanup efforts, and in the initial days immediately following the spill for which it was responsible, the first responses were actually initiated by local residents employing buckets, hoses, and skimmers. Overall funding for the cleanup was drastically limited because the Exxon Corporation was "unwilling to provide the hard cash," so in a mode of desperation, "the Federal Government brought in the US army's Corp of Engineers to provide assistance,"[135] who regularly quarreled with Exxon regarding the measures to be taken to clean up the oil.

As a political response to the disaster the US Congress passed the Oil Pollution Act of 1990, which prohibited oil transportation by any vessels that had previously spilled over one million gallons in any waterway, a move that in the grand scheme of things is analogous to applying a band-aid to a machine-gun wound, for is not one barrel of crude oil spilled carelessly in the ocean or along a majestic shoreline one too many? Considering these brief remarks concerning the fossil fuel industry, oil spills, and EPA policy decisions, it is all too obvious that such heedless actions have at their root a view of nature consistent with the types of secular *humanism* as analyzed, which emerges from and is grounded in the overarching *naturalistic* CF. However, as with unmitigated belief in the growth of science and technology, as encountered in Gray's critique of the *belief in human progress*, it appears that at least in the current political milieu (GOP/Republican Party) in the US, there is a prevalent *anti-intellectualism* and radical *anti-scientific* mindset. Here, we encounter a heightened form of anthropocentricism that harbors the radical type of *ethical egoism* embraced by Ayn Rand, where there is the conflation of a petty *selfishness* with glorified *self-interest*, and in this view, self-interest is linked with subject-hood as it is reduced and limited to the concern for the type of human happiness, or hedonism, grounded squarely in economic flourishing. And what we encounter in contemporary politics, at least as evidenced in the hard-line conservative far-right, which is now the dominant political force in the US since 2016 and the election of Trump, is *not* a concern for all people – or even with the type of priority given humans in the traditional understanding of anthropocentricism – but rather a concern for a narrow and select few, i.e., the wealthiest and whitest 1 percent of the populace. This current GOP movement, which it is not a stretch to call "nationalist," indeed this is Trump's own admitted and preferred moniker, lies far beyond the *neo-cons* of the 1990s and 2,000s. It is possible to understand US democracy in its current and radically transformed guise as instantiating a form of *anti-democracy*, in the form of either an *oligarchy* or a *plutocracy*, a state where the wealthiest are flagrantly exploiting *others* through the misuse and abuse of their power to pillage the environment in order to amass even greater fortunes, where their power is derived and inseparable from their wealth. It is even possible, in the extreme, to label the US government a *kleptocracy* due to the obscene levels of corruption that have been exposed as prevalent and ongoing within the Trump White House and among many of its executive cabinet members, and this includes, as mentioned above, members of the EPA. Here, we lose the sense of *normativity*, we abandon the stance embracing the solicitous care and guardianship that should be granted to nature – in terms of an ontological responsibility and an ethical response to the reticent ethical imperative such a responsibility conveys – which lies far beyond the mere concern for simply preserving the environment's resources along with the biological habitats of non-human entities.

Despite the *anti-intellectualism* prevalent in US politics, the type of volition we encounter is undoubtedly the anthropocentric *willing for* the selfish vested interests of the *few* and *willing against* the interests of the majority of the citizenry and *in opposition* to the preservation – not to mention the renewal of – the environment or nature. Here, nature, the *object-as-volitionally-encountered*, is identical to the description of nature we have already introduced as emerging from Cartesian and Kantian philosophical legacies, namely, nature conceived and encountered as nothing other than a massive usable, exploitable, and disposable resource-store for the selfish ends of human consumption. The *willing-for-the-sake-of-something-else* shows no concern for nature because its intrinsic value and meaning, as something radically other than an exploitable source for *economic* and *instrumental* gains of the ruling-classes, have been occluded. What we have introduced, despite reveling in a depraved and sadistic sense of cruelty, is most importantly about corporate production and profit and is comprised of and moved forward by, as stated above, massive layers of political greed and corruption. In the philosophy of Gray, as it relates to our talk of the *naturalistic* CF, this represents the contemporary evolution of *homo rapiens* as revealed in its perfect and crystalline, most ultra-predatory and destructive instantiation. Recall Leather stressing the need to change and replace a society grounded in the quest for profit with one focused more on the measured and controlled production of need, which bespeaks a transformed attunement toward nature. As stated, she found inspiration by drawing out the positive aspects of sustainable development, where we are concerned with an eco-phenomenological response to re-attune and change our societal worldview. Concluding with Toadvine and Brown, the "insights of eco-phenomenology hold promise of bringing about a dramatic shift in our current understanding of ourselves and our place in the natural world."[136] We note with great alarm that it is no longer the case that our ecological concern is *futural* in its direction, i.e., working to preserve the planet for future generations, for we are in the midst of a historical moment, a historical planetary crisis, where we have already inflicted what amounts to irreparable harm on the environment. We are living in a highly unique albeit frightening time where the potential exists of losing the *present moment* – not to mention the *future* – to global environmental destruction.

Notes

1 Michael Bonnett, "Normalizing the Catastrophe: Sustainability and Scientism," *Environmental Education Research*, 19, no. 1 (2013): 195.
2 Michael Bonnett, "Sustainability and Human Being," in *Post-Sustainability and Environmental Education*, eds. B. Jickling and S. Sterling (New York: Palgrave Macmillan, 2017), 3.
3 Lester Embree, "The Possibility of a Constitutive Phenomenology of the Environment," in *Eco-Phenomenology: Back to the Earth Itself*, eds. Charles Brown and Ted Toadvine (Albany: SUNY Press, 2003), 19–36.
4 Gabriel Marcel, *The Mystery of Being (Vol. I)*, trans. G. S. Fraser (New York: Gateway, 1959), 96.
5 www.activesustainability.com/environment/crimes-against-the-environment/.
6 John Gray, *Heresies: Against Progress and Other Illusions* (London: Granta Books, 2004), 43.
7 Sean Esbjorn-Hargens and Michael Zimmerman, *Integral Ecology: Uniting Multiple Perspectives on the Natural World* (Boston: Integral Books), 2009), 43.
8 Edmund Husserl, *The Crisis of European Sciences and Transcendental Phenomenology: An Introduction to Phenomenological Philosophy*, trans. David Carr (Illinois: Northwestern University Press, 1970).
9 Erazin Kohak, "An Understanding Heart: Reason, Value, and Transcendental Phenomenology," in *Eco-Phenomenology: Back to the Earth Itself*, eds. Charles Brown and Ted Toadvine (Albany: SUNY Press, 2003), 26.
10 Gray, *Heresies*, 115–123.
11 Kohak, "An Understanding Heart," 26.
12 Edmund Husserl, *The Basic Problems of Phenomenology: From the Lectures Winter Semester 1910–1911* (Dordrecht: Springer Press, 2006), 3.
13 Shaun Gallagher, *Phenomenology* (New York: Palgrave Macmillan, 2012), 159.
14 Dan Zahavi, *Husserl's Phenomenology* (Stanford: Stanford University Press, 2008), 125–127.
15 Dan Zahavi, "The Phenomenological Tradition," in *Routledge Companion to Twentieth Century Philosophy*, ed. D. Moran (London: Routledge, 2008), 10.
16 Gallagher, *Phenomenology*, 99.
17 Zahavi, *Husserl's Phenomenology*, 44.
18 Husserl, *The Crisis of European Sciences*, 108–109.
19 Kohak, "An Understanding Heart," 26.
20 Husserl, *The Crisis of European Sciences*.
21 Gallagher, *Phenomenology*, 160.
22 Charles Brown and Ted Toadvine, "Eco-Phenomenology: An Introduction," in *Eco-Phenomenology: Back to the Earth Itself*, eds. Charles Brown and Ted Toadvine (Albany: SUNY Press, 2003), xiii.
23 Michael Bonnett "The Powers that Be: Environmental Education and the Transcendent," *Policy Futures in Education*, 13, no. 1 (2015): 2.
24 Shaun Gallagher and Dan Zahavi, *The Phenomenological Mind* (New York: Routledge, 2012), 24.
25 Ibid.
26 Ibid., 25.
27 Ibid.
28 Martin Heidegger, *The Question Concerning Technology and Other Essays*, trans. William Lovitt (New York: Harper Torch Books, 1977).
29 Iain Thomson, "Environmental Philosophy," in *A Companion to Phenomenology and Existentialism*, eds. Mark Wrathall and Hubert Dreyfus (London: Wiley-Blackwell, 2009).

30 Martin Heidegger, *Discourse on Thinking*, trans. John Anderson and E. Hans Freund (New York: Harper-Perennial Books, 1969), 54.
31 Monika Langer, "Nietzsche, Heidegger, and Merleau-Ponty: Some of Their Contributions and Limitations for 'Environmentalism,'" in *Eco-Phenomenology: Back to the Earth Itself*, eds. Charles Brown and Ted Toadvine (Albany: SUNY Press, 2003), 112–113.
32 Michael Zimmerman, "Heidegger, Buddhism, and Deep Ecology," in *The Cambridge Companion to Heidegger*, ed. Charles Guignon (Cambridge: Cambridge University Press, 1993), 264.
33 Ibid.
34 Martin Heidegger, *Contributions to Philosophy (From Enowning)*, trans. K. Maly and P. Emad (Bloomington: Indiana University Press, 1999).
35 Martin Heidegger, "The Origin of the Work of Art," in *Basic Writings*, ed. David F. Krell (San Francisco: Harper-Collins, 1979), 201.
36 Ibid.
37 Martin Heidegger, *On Time and Being*, trans. Joan Stambaugh (New York: Harper & Row, 1972), 9.
38 Michael Zimmerman, *Heidegger's Confrontation with Modernity: Technology, Politics, Art* (Bloomington: Indiana University Press, 1990), 167.
39 Ibid., 168.
40 Martin Heidegger, *The History of Beyng*, trans. William McNeill and Jeffery Powell (Bloomington: Indiana University Press, 2015), 7.
41 Heidegger, *Question Concerning Technology*, 10.
42 Martin Heidegger, *Holderlin's Hymn "The Ister"*, trans. William McNeill and Julia Davis (Bloomington: Indiana University Press, 1996).
43 Heidegger, *Question Concerning Technology*, 11.
44 Zimmerman, *Heidegger's Confrontation with Modernity*, 169.
45 Heidegger, "Origin of the Work of Art," 189.
46 James Magrini and Elias Schwieler, *Reading Heidegger After the "Turn": At the Limits of Metaphysics* (New York: Routledge, 2017).
47 Vincent Scully, *The Earth, the Temple, and the Gods: Greek Sacred Architecture* (New Haven: Yale University Press, 1979), 3.
48 Heidegger, "Origin of the Work of Art," 149.
49 Ibid.
50 Ibid., 150.
51 Heidegger, *Question Concerning Technology*, 7.
52 Zimmerman, *Heidegger's Confrontation with Modernity*, 170.
53 Ibid.
54 Ibid., 7–8.
55 Ibid., 171.
56 Genesis 1:25–31 (Revised Standard Version).
57 Martin Heidegger, *Introduction to Metaphysics*, trans. Michael Fried and Richard Polt (New Haven: Yale University Press, 2001), 155.
58 Ibid., 207.
59 Ibid.
60 Zimmerman, *Heidegger's Confrontation with Modernity*, 182.
61 Ibid.
62 Heidegger, *History of Beyng*, 51–53.
63 Heidegger, *Question Concerning Technology*, 12.
64 Ecotricity, "The End of Fossil Fuels": www.ecotricity.co.uk.
65 Heidegger, *Question Concerning Technology*, 15.
66 Zimmerman, *Heidegger's Confrontation with Modernity*, 242.
67 Bonnett, "The Powers that Be," 2.
68 Ibid.

69 Michael Bonnett, *Retrieving Nature* (UK: Wiley-Blackwell, 2004), 159.

70 Raymond Tallis, *Aping Mankind: Neuromania, Darwinitis, and Misrepresentation of Humanity* (Durham: Acumen, 2011), 3.

71 Simon Critchley, "Passive Nihilism," *The Philosopher's Magazine*, 50, no. 1 (2010): 36–37.

72 John Gray, *Straw Dogs* (New York: Farrar, Straus, and Giroux, 2003), 4.

73 John Gray, *Heresies*, (London: Granta Books, 2004), 42.

74 Ibid., 43.

75 Ibid.

76 Jean-Paul Sartre, "Existentialism as a Humanism," in *Jean-Paul Sartre: Essays in Existentialism*, trans. Wade Baskin (New York: Citadel Press, 1995), 40.

77 Ibid.

78 Ibid.

79 John Gray, *Black Mass: Apocalyptic Religion and the Death of Utopia* (New York: Farrar, Straus, and Giroux, 2007), 44.

80 Ibid., 65

81 David Sprintzen, *Camus: A Critical Examination* (Philadelphia: Temple University Press, 1998), 166.

82 Ibid.

83 Gray, *Heresies*, 64

84 Ibid., 65

85 Tallis, *Aping Mankind*, 15.

86 David Detmer, *Phenomenology Explained: From Experience to Insight* (Chicago: Open Court, 2015), 64.

87 Bonnett, "Normalizing the Catastrophe," 192.

88 Charles Brown, "The Real and the Good," in *Eco-Phenomenology: Back to the Earth Itself*, eds. Charles Brown and Ted Toadvine (Albany: SUNY Press, 2003), 7.

89 Embree, "The Search for Immediacy and the Problem of the Political Life in Existentialism and Phenomenology," 40.

90 Ibid.

91 Ibid., 48.

92 Simon Critchley, *Continental Philosophy* (London: Oxford University Press, 2001), 17.

93 Ibid.

94 Ibid., 19.

95 Immanuel Kant, *Critique of Pure Reason*, trans. J. M. D. Meiklejohn (New York: St. Martin's Press, 1958), 22.

96 David Jardine, *To Dwell with a Boundless Heart* (New York: Peter Lang, 1998), 106.

97 Ibid., 108–109.

98 Ibid., 112.

99 Ibid.

100 Bonnett, "Normalizing the Catastrophe," 191.

101 Michael Bonnett, *Retrieving Nature: Education for a Post-Humanist Age* (London: Wiley-Blackwell, 2004), 114.

102 Bonnett, "Normalizing the Catastrophe," 192.

103 Kerry Turner, "Sustainability, Resource Conservation and Pollution Control: An Overview," in *Sustainable Environmental Management*, ed. Kerry Turner (London: Belhaven Press, 1998), 13.

104 Gray, *Heresies*, 121.

105 Michael Bonnett, "Environmental Education and the Issue of Nature," *Journal of Curriculum Studies*, 39, no. 6 (2007), 710.

106 Ibid.

107 Gray, *Straw Dogs*, 23.

108 Ibid., 17.

109 Ibid.
110 Ibid., 37.
111 Brown and Toadvine, "Eco-Phenomenology: An Introduction," xiii.
112 Gray, *Heresies*, 114, 115.
113 Ibid., 113.
114 Michael Gillespie, "The Search for Immediacy and the Problem of the Political Life in Existentialism and Phenomenology," in A *Companion to Phenomenology and Existentialism*, eds. Mark Wrathall and Hubert Dreyfus (London: Wiley-Blackwell, 2009), 536.
115 Ibid., 537.
116 Embree, "The Possibility of a Constitutive Phenomenology of the Environment," 47.
117 Brown, "The Real and the Good," 7.
118 Gillespie, "The Search for Immediacy and the Problem of the Political Life in Existentialism and Phenomenology," 538.
119 Ibid.
120 Ibid., 544.
121 Embree, "The Possibility of a Constitutive Phenomenology of the Environment," 46.
122 Amy Leather, "Hopelessly Devoted to Fossil Fuels," *Socialist Review* (2017): http://socialistreview.org.uk/420/hopelessly-devoted-fossil-fuels.
123 Barnini Chakraborty, "Paris Agreement on Climate Change: US Withdraws as Trump Calls it 'Unfair,'" (2017): www.foxnews.com.
124 Coral Davenport, "How Andrew Wheeler, the New Acting EPA Chief, Differs from Scott Pruitt," (2018): http://nytimes.com.
125 Nadja Popovich and Tatiana Schlossberg, "23 Environmental Rules Rolled Back in Trump's First 100 Days": www.nytimes.com/interactive/2017/05/02/climate/environmental-rules-reversed-trump-100-days.html.
126 Leather, "Hopelessly Devoted to Fossil Fuels."
127 Environmental Protection Agency (EPA), Memorandum: Improving EPA review of Appalachian surface coal mining operations under the Clean Water Act, National Environmental Policy Act, and the Environmental Justice Executive Order (Washington, DC, 2010), available at: www.uky.edu/CommInfoStudies/IRJCI/MTR guidance.pdf, accessed on April 1, 2019.
128 Melissa Horton, "What are the Effects of Fracking on the Environment?" (2017): www.investopedia.com.
129 Union of Concerned Scientists, "The Hidden Costs of Fossil Fuels: The true costs of coal, natural gas, and other fossil fuels aren't always obvious—but their impacts can be disastrous": www.ucsusa.org/clean-energy/coal-and-other-fossil-fuels/hidden-cost-of-fossils#bf-toc-0.
130 Megan Gannon, "Exxon Valdez 25th Anniversary: 5 Facts About the Historic Spill," (2014): www.livescience.com.
131 David Williamson, UNC News Services, "Exxon Valdez Oil Spill Effects Lasting far Longer than Expected," (2003, University of North Carolina at Chapel Hill).
132 Daniel Read, "Exxon Valdez: Corporate Greed and Environmental Catastrophe" (2009): www.marxism.com.
133 Greg Palast, "Court Rewards Exxon for Valdez Oil Spill," (2008): www.gregpalst.com.
134 Read, "Exxon Valdez: Corporate Greed and Environmental Catastrophe."
135 Ibid.
136 Brown and Toadvine, "Eco-Phenomenology: An Introduction," xiii.

Part II

The ethical call of *Nature*

Reticent imperatives

3 A reconceived philosophy of "objects" and *Nature*

Object-Oriented Ontology and eco-phenomenology

Toward a reconceived view of *nature*

Prior to exploring Harman's *Object Oriented Ontology* (OOO) in detail as it might be related to developing a conception of nature, we pause to briefly consider what Zimmerman and Esbjorn-Hargens have to say about the problems associated with modernity's conception of "nature" and the manner in which it influences the way we not only conceive of the natural world but more importantly, as is a running theme, how it is that the way we view nature – or are attuned to it – influences the manner of our dwelling within nature – which is crucial as we move forward, because in Chapter 4 we tackle the issues concerned with values and moral judgment. Modern science, according to Zimmerman and Esbjorn-Hargens, views nature in a reductive manner, as a "perfectly harmonious and interrelated system," in terms of a great "interlocking order of beings, each mutually interwoven with all others, but wholly lacking interiority and intersubjectivity."[1] Modern science is skeptical about the idea of *interiorities* and tends to privilege *exteriors* to the extreme of relegating the idea of nature's *interiority* to "the same category as other discredited ideas, such as the soul."[2] This is consistent with a *materialist* view of nature, and there are, as the two authors contend, in line with our eco-phenomenological view, problems associated with attempting to defend this view of the world, not the least of which is that *normativity* is excluded from a material view of the natural world, "because value is not a material object for study."[3] To move beyond this limited view, the authors suggest an understanding of the natural world that has three distinct manifestations that are interwoven and manifest in different ways, appearing from uniquely diverging perspectives, and these views are given the following designations: NATURE, Nature, and nature, the first referring to the "Great Nest of Being," the second naming the "Great Web of Life," and the third categorizing the "Great Biosphere." We will now take the terms each in turn in an effort to clarify them.

"NATURE" is representative of the entire cosmos, which includes all of its interior and exterior dimensions, and so includes the dimensions termed "Nature" and "nature." Of these two latter terms, the first refers to the exterior dimensions of experience accessible to the natural sciences and social sciences.

The second term refers to the empirical-sensate world of experience, consisting at once of "exterior world disclosed by the five senses (and their extensions), and the interior world disclosed by feelings, emotional-sexual impulses, somatic experiences as contrasted with rational mind and culture,"[4] and it is this realm, which is also expressive of "spiritual" experiences, which is linked to phenomenology by the authors. So, to further clarify these experiential dimensions, the authors ask readers to consider the following three scenarios:

1 When we find ourselves in the midst of the natural world, and in a moment of ecstatic transfiguration, attuned to the wonder (*thaumazein*) of all that is around, we exclaim, "Wow! Nature is sublime and wondrous!" we are experiencing NATURE, and we add, as will be developed in Chapters 4 and 5, that this dimension bespeaks what we identify as nature's integrity and its intrinsic value;
2 When encountering the surfaces or exteriors of the entire empirical-sensory world, we utter, "My! Isn't nature beautiful?" we are having the experience associated with Nature. We must note that this designation also includes the view of the natural world consistent with the empirical sciences; and finally,
3 When we experience the natural world in terms that are momentarily beyond our rational – or *naturalistic* – and culturally constructed responses to it, e.g., when we are out on a nature hike and observe and state, "Hiking allows me to clear my head and get in touch with nature," we are experiencing nature.[5]

As stated, phenomenological inquiry as linked to eco-phenomenology is related to the dimension of experience termed "nature," with its interior potential for emotional and spiritual development in relation to what is identified as an ontological-epistemological relationship to the natural world, and indeed, the authors claim that as our interior dimension develops and flourishes, as a sense of "genuine environmental ethic" becomes *as* possibility, we become "worldcentric" with the potential for becoming "planet-centric."[6] We add that it is also possible to imagine the pre-reflective immersion in the world that phenomenology reveals, offering us the potential to make connections with the natural world that can be understood in terms of an experience of transcendence, a form of attunement that would be consistent with the transformative mindset described by the authors in relation to NATURE, and this is a view we also find in Neil Evernden's view of "nature-as-miracle" – or wondrous encounter – which might be conceived as a phenomenologically inspired experience that "transcends the normal understanding and holds it temporarily in abeyance so that the personal awareness of the living world is restored."[7] If we consider the interpretation offered by Esbjorn-Hargens and Zimmerman in relation to our reading, it is possible to conclude that phenomenology works to bypass the *naturalistic* attitude that would be associated with what they designate Nature, which as stated is a dimension concerned with the exterior dimensions of experience accessible to the natural sciences, representing but one dimension of NATURE. Importantly, as stated above, Esbjorn-Hargens and Zimmerman emphasize that in addition to the "interior" dimension of human beings, there is an undeniable interior dimension to the natural world. This foregoing line of thought might be said to intimate eco-phenomenology's concern with the problems and difficulties haunting the attempt to reveal and cultivate an authentic view of NATURE, this due to the dominance of views emerging from and traceable to the natural sciences, giving rise to and intensifying the ever-present pervasiveness of the *naturalistic* attitude that colors and attunes our everyday modes of world-disclosure, highlighted by the presence of Nature, insidiously working to obscure a *phenomenological* view toward NATURE. For we must continually be aware that the exclusive reliance on the view from science forecloses the phenomenological dimension of experience that the authors identify, because science,

> however valuable in its own right, provides no account for the interiority of natural phenomena, much less for the interiority of human beings. Such interiority has no "place" in a world composed solely of material particles and complex systems.... It only worships objects out there [exteriority], and ignores the subjectivity [interiority] in here.[8]

Let us pause briefly to consider the human's relationship with *Nature* and *vice versa* that might be drawn from the initial ground covered, with the understanding that this view will be deepened, problematized, and amended as we proceed forward in the analysis. Here, the reader will note that we are employing italics and capitalization when referencing "nature" as "*Nature*," and we explain this typographical choice in the final paragraph of this section. We've already discussed the problems with Deep Ecology in Chapter 1 and we want to return to its view of the human-and-nature that it endorses. As Toadvine recognizes, Deep Ecology moves toward the "popular environmentalism of nature-as-self, the view that postulates a kinship relation between humans and the larger eco-system."[9] The problem, as Toadvine observes, is that this view is no less an instance of the "dialectic construction" we encounter in the "Nature-as-object" view, which is why we must be aware and suspicious of "the anthropomorphic and even Romantic overtones of a return to holistic oneness with nature conceived as an organized and productive telic system."[10] Don Marietta claims that such *holistic* pictures arise from "several misuses of concepts that were developed as conceptual models by biologists,"[11] and he identifies these ways of misunderstanding *Nature* as *exaggeration*, *reductionism*, and *excessive abstraction*. Exaggeration is expressed in the tendency, without substantial backing evidence, in a grand and hyperbolic manner, to inflate the interrelatedness of things in the system of *Nature*. Reductionism occurs when we fail to take into account the deep complexity and inconceivable magnitude of not only the natural world, but also the human being. We encounter abstractionism when our modes for representing the human-and-nature relationship result from us projecting on

Nature a *thematized* – in terms of the universal (*sub specie aeternitatis*) – view that is "far beyond the actual data about organisms in a natural system and the ways they are interrelated."[12]

As introduced, Deep Ecology adopts a *relational total field* model of the human-and-nature, which expresses that the "intrinsic relation between two things A and B is such that the relation belongs to the definitions or basic constitutions of A and B, so that without the relation, A and B are no longer the same things."[13] Based on our discussion thus far, it is obvious that this concept-as-model is flawed. For the *relational field* model is suggesting, in a highly disingenuous manner, that there is an intimate, and beyond, an undeniable and indelible symbiotic relationship between the human and *Nature*, a relationship that might be expressed in terms of what we shall call "universal mutualism." This view indicates a close relationship between two organisms wherein each organism benefits and they are in fact each dependent on the other, and we might say, in such a way that they are inextricably co-dependent. For example, *mutualism* is instantiated within the relationship between the human being and microorganisms, as evident, for instance, in the metabolic processes of digestion. Microorganisms in the gastro-intestinal tract are necessary to ensure proper digestion, while the environment of the human's digestive organs facilitates the flourishing of the microorganisms. Although this and many other examples of *mutualism* occur on what might be called a *local* or *particular* level, it is at once an instance of *exaggeration*, *reductionism*, and *abstraction* to generalize and then posit a binding and indelible *universal* model of symbiosis in the form of *mutualism* that defines the human-and-*Nature* relationship *in toto*; symbiosis is not universal in this "strong" sense. For example, in the symbiotic model of *commensalism*, only one of the organisms in the relationship benefits, while the other neither benefits nor suffers.

We might also consider another form of symbiosis, *parasitism*, because importantly this is also an inappropriate model for the human-and-*Nature* relationship that eco-phenomenology embraces. It is not the case that in relation to the human being the Earth/*Nature* manifests or presences as *host* and the human both exists and subsists as a *parasite*, feeding, growing, and habitually taking advantage of *Nature* while in turn contributing nothing to its continuing survival, and in fact, hastening its decline and ultimate eradication. We address the issue of symbiosis in order to understand it, in terms of Harman's OOO, as potentially *non-reciprocal* in nature or "*asymmetrical*."[14] One way to begin thinking about the human-and-*Nature* is in terms of the debt we owe to the Earth or *Nature* for our survival and continued thriving, *Nature*, we might say, borrowing from the language of the medieval theological tradition concerned specifically with the "proofs" for God's existence – as in St. Aquinas's "five ways" (*Summa Theologica*) – instantiates both the cause *in fieri* and cause *in esse* of humans and non-human species and life-forms that populate and inhabit the planet. The first Latin term, as philosopher Paul Edwards informs us, refers to a "factor which brought or helped to bring an effect into existence."[15] The second Latin term refers to a factor that "sustains or helps to sustain the effect 'in being.'"[16] Moving forward we attempt to develop a view of the human-and-*Nature* that is suggested in part through Marietta's eco-phenomenology, which we will broaden and enhance when discussing and analyzing Harman's OOO. Marietta argues that the relationship between the human-and-*Nature* must be one where our ontology does not "exaggerate either unity or divergence," and we should be open to "accept the sense in which the world is 'out there' without thinking of the world as an alien system of which we are wholly independent."[17]

When employing the italicized term *Nature*, we include an understanding, which might be referred to as a *third view of Nature*, which draws inspiration from Harman's OOO. In the 2012 pamphlet *The Third Table*, which establishes the intimate relationship between philosophy and art – and we later speak in detail regarding Harman's claim that philosophy, like art, does not give us "knowledge" proper or "truth" – Harman analyzes and criticizes Arthur Stanley Eddington's metaphor of the *two tables*, where one table is experienced as a practical and usable object, and the other is experienced through a mode of world-disclosure that is scientific. Table No. 1 is "practical, solid, colored, hard," whereas table No. 2 is "made up of empty space swarming with electrons and other tiny particles and invisible fields."[18] The so-called "Third Table" is far more *mysterious*, *elusive*, and *alluring*, withdrawing from all interactions with both humans and other objects. Moving forward, we learn that the same might be said of our conception of *Nature*, where *nature one* is disingenuously reduced in knowledge to its instrumental or practical capacity as object for human mastery and manipulation, and *nature two* is disingenuously reducible to its smallest physical components, as is consistent with physics. However, much like the *third table* there is a *third Nature*, which is a "world filled primarily not with electrons or human praxis, but with ghostly objects withdrawing from human and inhuman access, accessible," not as an object of knowledge, but rather as a phenomenon experienced and approached obliquely through *allusion* and *intimation*, a force that brings us close to its essence by "seducing us by means of *allure*."[19] What we have described here briefly is a re-conceptualization of *Nature* that resists all human attempts to know it, which as stated, must be approached only *indirectly* through a form of philosophizing, which as we learn, incorporates description and metaphor, and so is relatable to our view of eco-phenomenology, a way of practicing philosophy that has more in common with aesthetics than science; and this is precisely the view that OOO embraces, and it is now to Harman's unique philosophy/ontology of objects that we turn.

A detailed overview of Graham Harman's *Object-Oriented Ontology*

The emergence of the reality of objects: On reinterpreting Husserl and Heidegger

As we learn moving into Chapter 4, the view of *Nature* that can be culled from Harman's unique philosophy is radically other than the views of nature we encountered in Thomson's definitions of *naturalistic* ethical realism and *transcendental* ethical realism. Recall that in the first view Thomson philosophized that issues of good and bad, ethical and unethical, are based on "facts" as related to an understanding of nature that is developed in terms of nature's universal traits. And in the second of these eco-phenomenological ethical views articulated by Thomson, he stresses the notion that nature lives in a way that resembles Heidegger's understanding of the Earth, which instantiates a relationship to nature through which we discover what is of value in terms that are neither reducible to nor consistent with facts and values from a traditional philosophical perspective. Due to the depth and breadth of Harman's continually developing philosophy, in light of his prolific scholarly output, it is impossible to offer anything resembling an all-encompassing elucidation and analysis of all the aspects of OOO in a single chapter. What we do, however, is focus on specific components of OOO that are relatable to our developing eco-phenomenological position. Thus, the contents of this chapter should not be thought of as an attempt to completely square OOO with our eco-phenomenological project – leaving no remainder, as it were – for any such move would not only require far more space than is allotted in this chapter, but would also, more than likely, prove extremely difficult if not impossible, for we must recall Harman's compelling critique of phenomenology and the limits of interpretation from Chapter 1. In what follows we identify the following areas of common agreement, while remaining true to, by engaging in a charitable reading of, Harman's ontology of objects, which might prove beneficial to our re-conceptualization and practice of eco-phenomenology: (1) The adamant rejection of *anthropocentrism*, which includes the critique of and move beyond a traditional metaphysical view of *correlationalism* in order to speak of a *realist* ontology of objects as external to the human being; (2) The rejection of both an *essentialist* and an *instrumental* view of objects that reduces them to either their material composition or their use and function; and (3) The power of metaphor in description to potentially clear an inroad through which to obliquely access the inner essence of objects, which always *withdraws* from our grasp, *withholding* its secret and inaccessible interior, viewed by Harman in highly descriptive terms as a "volcanic core," or "kaleidoscope, witch's cauldron, steel mill, or alchemist's flask."[20]

As is evident, we are critical of *metaphysical instrumentality*. It must be noted, however, that although we are adopting a critical view of the *metaphysics of presence*, Harman embraces an *ontological* view that he equates with a non-problematic understanding of "metaphysics," i.e., a view of metaphysics that stands beyond the *metaphysics of presence*. But this is not to indicate that Harman is uncritical of a certain disingenuous view and practice of "metaphysics," and his concern, as might be related to ours, is focused on what he identifies as a form of *metaphysics* equated with the predominant view of the human, world, and entities populating the world manifesting *as* "Continental philosophy." We have already discussed critiques of Kant, Descartes, Husserl, and Heidegger in relation to eco-phenomenology, but it is crucial that we review Harman's critique of Kant, Husserl, and Heidegger as related to OOO, and we learn that Harman's conclusions, although neither expressed nor incorporated in eco-phenomenological circles, relate to our concern with rethinking the human's relationship with *Nature*. To begin, in his attempt to get us to rethink the pejorative characterization of Western metaphysics in terms of the *metaphysics of presence*, Harman emphasizes two problems with Continental philosophy as related to the "serious effort to defend the credentials of an independent reality beyond appearance, of a substance beyond every series of qualities, of a world-in-itself in which the human subject plays just one limited part."[21] First, Harman is critical of the bias of the "philosophy of access," i.e., in which instead of approaching and discussing reality itself, philosophy must perform a "series of sophisticated critical and self-reflexive maneuvers so as to ensure that we are only talking about objects as they manifest themselves to us, not about objects in their own inner life."[22] Second, a related bias is identified as the "philosophy of contexts or networks," which views the notion of an independent substance as naïve, for what is most primary is the "totality of meanings or objects in their reciprocal relations; any individual part of this network is nothing but an abstraction broken away from the system as a whole."[23] According to Harman, Kant's philosophy expresses the precise *type* of relation that plays a central role within the Western philosophical tradition, namely, the division between human (mind) and world (matter). This represents *correlationism*, which locates the conditions for the possibility of all experience within the context of or the "correlation" between the human mind or consciousness and the world, or as it is more commonly understood, the correlation between thought and Being. This relationship between mind and world is understood as occurring "in space and time and in terms of the twelve different categories of the understanding";[24] thus, to reiterate, the condition of possibility for all empirical experience is *mind-dependent*, embodied in the *transcendental subject*. We note here that there is another significant idea from Kant's philosophy that is of concern, namely, the issue of the *noumena* or "things-in-themselves" that Harman appropriates and reinterprets, which we discuss in this chapter.

Harman is critical of Husserl's phenomenology because it discounts the independent existence of objects, for "Husserl is unconcerned with the reality of things outside their accessibility to consciousness."[25] However, Harman makes clear that objects do play a crucial role in Husserl's phenomenology, which places "stress on the object rather than its qualities,"[26] this because of the shifting and variable nature of an object's qualities. Indeed, for Husserl the human's conscious experience is not primarily about the "contents" of its thoughts but rather about the "objects" that it intends. Thus, as Harman asserts, despite the "bracketing of the natural world, Husserl does not reduce objects to our experience of objects,"[27] and because of this tendency he does, in a manner that went unrecognized to Husserl, restore objects to the "throne in the form of intentional objects,"[28] i.e., the latent presence of objects subtends intentional objects – and, as we move to Heidegger, it is possible to state that in both cases "a *real object* silently labours in the depths of their philosophies."[29]

With respect to Heidegger, we must pause to take a closer look at Harman's unique – and some would say controversial – interpretation, for this reading is the key focus of Harman's early texts introducing a radical new philosophy or metaphysics of objects. It is crucial to recognize that Harman's reading of *Being and Time* and Heidegger's phenomenology of *tool-Being* (Dasein and *equipmentality*) is radically opposed to typical – and we might say canonical – readings that focus on privileging *Zuhandenheit* ("readiness-to-hand") over *Vorhandenheit* ("presence-at-hand") and thus the realm of Dasein's *practical* comportment *in* the world over its *theoretical* abstractions *of* the world, and in the extreme, some readings emerging from this tradition that align Heidegger with *pragmatism*. Let us explore the implications of Harman's unique reading of Heidegger. To begin, in the traditional Heideggerian reading of *tool-Being*, "presence-at-hand" (*Vorhandenheit*) is indicative of that which is "independent of Dasein," and *presence* always fails to do "justice to a thing's full reality, which withdraws into a depth no awareness can ever exhaust."[30] One important way things are made present is through scientific theorizing, "whereby we view (and thereby reduce) entities [to] pieces of objective matter occupying time and space," possessing properties that can be categorized, explained, and known. It is wrongly assumed, as Harman reveals, that because "present-at-hand" demonstrates an independence from Dasein, and further, since Heidegger is highly critical of "presence-at-hand," this by extension serves as a critique of "realism" and therefore establishes Heidegger as espousing an *anti-realist* position. With this misunderstanding, it is concluded that what is present-at-hand is *independent of Dasein* and what is ready-to-hand is *dependent on Dasein*, i.e., items of "equipment ultimately [gain] meaning only from that 'for the sake of which' they are used: namely, for the sake of Dasein's own existence."[31]

Harman corrects and reverses this understanding and then provides an interpretation that is truly unique in the annals of Continental and Analytic (*pragmatist*) Heideggerian scholarship. His initial claim is that "Heidegger's presence-at-hand means dependence on Dasein," as in scientific explanations that project mathematical schemas for understanding nature, always present "*for* someone or something," and "readiness-to-hand means the independence of that which withdraws from access."[32] At first blush, it appears that it is only *Zuhandenheit* that represents the type of independence from Dasein that is indicative of a *withdrawal* of an understanding of objects in *use*. However, Harman assures us that in both realms – and here we include *Vorhandenheit* – objects *withdraw* from the human's full understanding of them, for "unconscious practice distorts or translates the things of the world no less than conscious theory does,"[33] and thus it is concluded that we can never arrive at a complete understanding of the essential reality of things. The invisible use of a tool or manipulation of an object does not exhaust its reality; neither does a description of its properties or a scientific analysis of its anatomic structure, for there is always an *excess* of that which is *held-back* in all of our encounters with objects, which includes both our *practical* and our *theoretical* dealings and encounters with them. What Heidegger gives us, according to Harman, is a phenomenological-ontology of objects rather than a strict phenomenology of Dasein's practical comportment (ready-to-hand) unfolding in terms of its *relational totality* or its *world*. Heidegger emerges from this analysis as a "realist," as a philosopher who acknowledges the existence of a mind-independent reality, for despite Heidegger prioritizing *aletheia* above the Correspondence Model of Truth, which is "the most common form of truth found among realists,"[34] this does not amount to evidence defending Heidegger's rejection "of realism *tout court*," for it is "possible to believe in a mind-independent reality while not believing it possible to attain a perfectly lucid grasp of the reality."[35] Indeed, within an extremely short passage in the now-famous §6, ¶44 ("Dasein, Disclosedness, and Truth") of *Being and Time*, Heidegger is unambiguous regarding the existence of a Dasein-independent reality when he reasons that prior to Newton's laws being discovered, they were neither true nor false, but to say that these laws were neither true nor false before Newton is *not* the equivalent of claiming that such laws did not exist or obtain, that there "were no such entities as have been uncovered and pointed out by those laws," as an integral phenomenon of a world and a reality external to Dasein.[36]

Philosophical realism or the ontology of objects

Harman's ontology of objects reflects a new form of "Continental realism," for as we know, according to Harman, Continental philosophy has adopted an *anti-realist* perspective. It is possible to define OOO in the following terms, and we should recognize that it too, in its own unique manner, like eco-phenomenology, seeks to access, however obliquely, the secret and hidden world of *relationships* between objects, which includes a concern for the bonds or tensions between objects and their qualities, lying beneath the modes of *everyday* cognition and the *naturalistic* view of objects. As opposed to *direct realism*, OOO is a unique and complex form of *realism* espousing a reality that is independent of the human mind and consciousness, which is approached through a manner of philosophizing that is more reminiscent of aesthetics than science, which *indirectly* reveals a view of objects in terms of a "darker form of 'weird realism' bearing little resemblance to the presuppositions of everyday life."[37] We must stress that *realism* here is not to be equated with *materialism*, which "idealizes its objects by reducing them to a limited number of mathematizable features expressed by the accidental state of present day physics."[38] In relation to our concerns as drawn out thus far, Harman's ontology is radically anti-*anthropocentric* in at least two ways: (1) Humans as objects are on par with every other object that populates the universe and (2) Human knowledge, as opposed to the view embraced by strains of secular humanism discussed earlier, is not merely limited, it is *impossible*, for "no one is actually in possession of knowledge or truth," where truth is understood in terms of categorical knowing, and Harman observes that the real danger to thought is "not relativism but *idealism*," for no "theory survives its first encounter with reality."[39] Thus, there is not only a de-centering occurring with respect to the human subject, but there is also a flattening and leveling of the human subject when we consider that "the vast majority of relations in the universe do not involve human beings,"[40] and that objects are continually interacting with each other in the absence of humans;

this, states Harman, is what a "pro-object theory" requires, namely, the awareness that relationships between objects do not and need not directly involve humans. Reality, in terms of objects and their relations, *withdraws* from full disclosure and *withholds* itself from all attempts to know it, and this notion of reality- or *object-withdrawal*, which is an expression of ontological "limitations" (*finitude*), is unique to Harman's OOO. For he argues that it is not only that objects withdraw from human knowledge, but that objects also withdraw from each other, and this represents a central tenet of OOO as well as a central feature of objects and their relations. This is because Harman extends the concept of *finitude* beyond the human condition, for finitude does not "haunt human beings alone"; rather, the "tragic burden of finitude is shouldered" by all objects.[41]

As might be related to our critique of secular humanism and the pernicious form of environmental philosophy emerging from it, as grounded in a CF given structure by *scientism*, Harman is also highly critical of the various ways in which we distort, caricature, and obscure "reality" when we attempt to *exhaust it in knowledge* through a variety of means, methods, and disciplines. As introduced in Chapter 1, Harman claims that we *undermine* objects when we attempt to explain them in terms of their "smaller constituent, by way of a downward reduction," which renders objects superficial, relegating them to an inconsequential or epi-phenomenological status.[42] We *overmine* objects when we treat them as "needlessly deep or spooky hypotheses by comparison with their tangible properties or effects," e.g., as in the empiricist *bundle theory of objects*, which holds that objects are illusory, they are nothing beyond their qualities or the effects they enact. However, rarely do we encounter these two aforementioned forms of "knowledge" in isolation, rather they occur in the phenomenon of *duomining*, and here we should recall the critique of *scientism* from Chapter 2, for *duomining* occurs in scientific materialism, which on the one hand *undermines* objects by viewing them as "ultimate particles, fields, strings, or indeterminate 'matter' as the ultimate layer of the cosmos," and on the other, *overmines* objects by making the claim that "mathematics can exhaust the primary qualities of the genuine layer."[43] An object can never be reduced either in a downward or upward direction, neither to its composition or constitution nor to the effects or use that it lends itself to, i.e., all objects resist being exhausted by *undermining*, *overmining*, or *duomining*. What Harman refers to as the "mutual darkness of objects"[44] indicates that objects instantiate a *surplus* or withhold a *remainder* that is beyond its component parts and beneath the totality of its effects, and for this reason, we can have no categorical "knowledge" of or "truth" about objects. This view is also shared by Lingis, who when writing on "synesthetic perception," emphasizes that an object, such as a table, "is not so many impressions imprinted on our surfaces nor the sum total of its functional uses," but rather, and more importantly, "it is contained in itself, exists beyond all we could ever itemize off it."[45]

What, according to OOO, is an object? How might we begin to understand it? The history of philosophy treats objects as if they were "entities that are inanimate, durable, non-human, or made of physical matter."[46] Harman employs the term object to mean anything at all that exists, i.e., *anything* that resists reduction to material constituent parts or the effects that it has on the world or other objects. Thus, pens, staplers, scissors, tables, chairs, bacteria, flowers, mountains, sand, and even unicorns and angels are all classified as objects. As opposed to *eidetic* or what we might call "imaginary" objects, material objects and their relations do not depend on human existence or intervention, and they all have an internal aspect that cannot be accessed by means of attending merely to an object's exterior. Objects are always antecedent to either their use or qualities, although they depend on those qualities. As Harman stresses, the "sun isn't 'used'; *it is*,"[47] and the same might be said of mountains, sprawling meadows, rivers, and non-human sentient creatures – in short, *Nature* is not used; *it is*. Beyond this, every real event is also considered an object despite its endurance or duration: Harman provides the example of a mid-air plane collision – a collision of two Real Objects (RO) – which instantiates because it is the origin of another object, the "collision object" or *collision-event*, of which the planes, however momentarily, are in the interior. The collision and destruction of the planes are the two component objects comprising the "collision-object," which produces the "retroactive effects on its parts of a larger collision entity, to which we never pay attention because it lasts so briefly and takes little or no physical form."[48] For in the seconds after the collision, we experience the "decomposition of the collision-object into its initial components."[49] As we might relate to our example of the Exxon Valdez disaster (EVD), we encounter a RO in the tanker carrying crude oil in its relation to the coastline RO, which produces a third object that might be termed the "disaster-object," and indeed it is possible to state that Global Warming is really a number of different objects, which are continually in the process of being created and then disintegrating. Moving forward, in line with the principles of OOO and the characteristics that objects display, we offer the following analysis:

1 All objects should receive equal attention, and this is why Harman embraces a *flat ontology* for casting a wide net that includes far more of the world in its analyses than merely focusing on the dual components of *correlationism*; for Harman, good philosophy tells us "about the features that belong to *everything*";[50]
2 Objects and their properties are not identical and there exists a tension in their relationship that initiates change;
3 Objects are either *real* or *sensual* and real objects exist even if they exact no influence on other objects, whereas sensual objects exist strictly in relation to real objects. Real objects neither relate directly to nor interact with other real objects, but only share a relationship through the mediation of a sensual object and sensual qualities; and
4 There exist two kinds of objects and two kinds of sensual qualities that form a quadruplicity, as understood in terms of OOO's *ontography*, through which four specific relations, interactions, or "tensions," namely, *space*, *time*, *eidos*, and *essence*, are instantiated.

The quadruplicity of objects

As we have introduced, OOO's ontography focuses on the "rifts between the two kinds of objects and their two kinds of qualities."[51] This ontography is represented within a quadruple schema (*flat ontology*) consisting of objects and their relations; we might say that this represents the crucial distinction between what *presents* and what *is*: (1) Real Objects (RO), (2) Real Qualities (RQ), (3) Sensual Objects (SO), and Sensual Qualities (SQ). A brief explanation of objects and qualities is in order. Real objects exist whether or not there are sentient entities present to cognize them; thus to reiterate, they do not depend on human beings or non-human animals. Harman is clear that "sensual objects exist only as the correlate of our acts of consciousness,"[52] so we do not determine either SOs or SQs, rather we experience them in and through our interaction with real objects. This notion flies in the face of empiricism and the view that privileges the human subject as the conscious "creator" of the world, for there is "no pre-given blur of sense data that then gets molded into tangible units."[53] As stated, OOO seeks to graduate beyond the ontological correlation composed of mind and world, which in great part harbors the belief that humans determine the manner in which their world shows up. Harman goes on to make the more radical claim that not only is it the case that humans do not determine objects, *objects themselves* do not determine other objects, and to continue this line of thought, just as objects always remain opaque to human beings, "objects are opaque *to each other*,"[54] i.e., although objects interact with each other, the "internal information of each remains hidden (or 'encapsulated') from the others."[55] Every object shelters its essence from all other objects, and in short, objects remain mysterious to other objects. What is obvious here is that OOO rejects any form of *metaphysical determinism*, but what is slightly less obvious is that a philosophy of *metaphysical indeterminism* is operative below the surface of the analysis, for OOO rejects the attempt to understand objects through the type of description, explanation, and prediction necessary for scientific explanation, for instance as related to explaining antecedent cause and effect. What we instead encounter is a view of objects in which their behavior can neither be *predicted* nor *known* with categorical certainty based on the inherent ontological structure of the universe, where chance, uncertainty, and randomness break open the possibility, to borrow a phrase from Aristotle, which we return to in Chapter 4, that *things can always be otherwise, other than they are*.

Based on the quadruple schema presented in Harman's OOO, we encounter specific interactions or *tensions* between the four poles (*object/qualities rift*) of RO, RQ, SO, SQ as follows:

1. The RO⇒SQ tension, which, for Harman, is indicative of *space*, referring to the distance and proximity between real objects and sensual qualities;
2. The SO⇒SQ tension, which is indicative of *time*, in that sensual qualities change over time as the SO gives off or presents different qualities in a *perspectival* manner, while the SO endures;

3 The SO⇒RQ tension, which is indicative of the *eidos*, for the SO is traceable to the RQ that exists whether or not we are aware of them; and
4 The RO⇒RQ tension, which indicates that real qualities belong to real objects in terms of *essence*, although no knowledge can be had of *essence* through either sensual or intellectual intuition.

Much in the way that we have redefined "essence" in Chapter 1 in relation to traditional metaphysics, Harman understands "essence" in terms radically other than post-modernism, which embraces a view of "essence" as "politically repressive"; but for Harman, in line with our view, "essence" is not the equivalent of an indelible, *hypostatized* inner core of a thing that can be gleaned or known through intellectual intuition, it is not something – indeed it is not a "thing" at all – that anyone can ever claim to have *knowledge* of.

We now must consider the manner in which it is possible for objects to influence each other, considering, as introduced, that objects "exceed the grasp not only of all human theory, perception and practical action, but of every sort of direct relation"[56] with other objects, as they are incapable of coming into direct contact with each other. Harman offers a unique and complex line of argumentation for the way in which objects "*touch without touching*, through some sort of indirect contact," and this is what OOO identifies as "vicarious causation," which we explore below.

Harman claims that the "primary meaning of 'cause' is to create a new object," and only secondarily does it mean that an "object has an effect on others or retrograde impact on its own parts."[57] This notion initially strikes us as strange, considering that we now know that the creation of a new object does not occur through direct contact between ROs; however, although it is argued that "objects withdraw into inscrutable depths," it is the case that "they somehow relate, or nothing would happen."[58] To begin to understand *causation* in OOO we turn to a theory of causation found in *occasionalism*, which manifests in the history of philosophy in both theological and secular forms: In the former, objects do not incite causation, they do not exercise their own power or rely on human power to bring "effects" into existence. Rather, it is God, acting as a mediator, who intervenes and is ultimately responsible for causation. For "God is not just the only creator," asserts Harman, "but the sole causal agent in the universe."[59] The latter and "secular" form of *occasionalism*, found in Hume and then later modified in Kant, holds that *one special entity in particular* serves as the locus of causation and this is the human mind. In a unique modification to this form of secular *occasionalism*, Harman claims that although ROs do not make direct contact, as in the case of direct impact, they do interact or are related by way of the "fictional images they present to each other,"[60] which indicates that the encounter between two ROs occurs only through the mediated connection established between the sensuous versions or presentations of each object – here, recall the quadruplicity of objects as introduced above – in such a way that "there are [potential] retroactive effects on the real,"[61] and this is *vicarious causation*. Thus, causation occurs between two ROs through the mediation of their sensual faces, or SOs and SQs, and in the process of causation what is revealed, "despite their uniting in a bond," amounts to distorted translations of the ROs involved, for as stated, even objects do not encounter the elusive *essence* or *thing in itself* of objects. The exceptionally radical and novel nature of this view of objects, causation, and the limited view of "knowledge" access occurs in Harman's extension of this idea beyond human beings – recall our critique of *correlationalism* – to include all objects, which ontologically represents the phenomenon of the essence of a RO *withdrawing* from all other ROs.

> The withdrawal of things from all access is not some quirky existential/psychological feature of humans, but infects even the most rudimentary forms of inanimate causation. Veiling and unveiling are ubiquitous: even between billiard balls, even between fire and cotton, and even when humans are not observing, do not yet exist, or exist no longer.[62]

With this view of causality and the interaction between the qualities of objects, OOO is certainly not advocating a philosophical version of *panpsychism*, a view wherein the material objects populating the world, no matter their status, possess in some manner or degree consciousness. Harman understands quite well that only a *charlatan* "would grant cognitive powers to flowers or sand," but as he suggests, "those powers are merely of a special case of what must be called 'relations more generally,'"[63] and in Harman's flat ontology such relationships are considered *intentional acts* even among inanimate entities, "since the *relata* necessarily relate to sensuous objects rather than real ones."[64] To reiterate, all relations between objects, based on the view of *vicarious causation*, produce a new object, which we have already referred to as an "object-event," and so my relation with a tree, mountain, or wild animal lives at the same causal level, so to speak, as a bird's or insect's interaction with the aforementioned objects. For in each case the RO (tree), in line with the essential structure of *vicarious causation*, withdraws as it presents its *sensual face* to and for the other, let us say, an insect, and when both interact within the interior of a new object this instantiates an intentional relation as a whole. This notion of encountering the *sensual face* of an object has important implications for our eco-phenomenological analysis as we learn in the forthcoming chapters, when we move into the phenomenology of Lingis and his philosophy of "imperatives" emanating from the *faces of the things of the world*. The phenomenon of causation, as philosophized by Harman, experienced by the human in its interaction with *Nature*, when its mode of Being-in-the-world is brought into close proximity to the *volcanic core* of objects – the *thing in itself* – results in a moment of attunement, the experience of an "effect," as discussed in Chapter 1, that accompanies the experience, which we link to Harman's understanding of *sincerity*; and this interpretation eventually leads us in Chapters 4 and 5 to an eco-phenomenological ethical analysis of the human being in its reconceived relation to *Nature*.

Harman intimates a view of the natural world, which is understood as an "object," that sits within the overarching ontology of objects (ontography) as discussed, where objects and the interaction of their qualities is primarily understood in terms of relationships that preserve the overarching mystery inherent in all things, in that objects will always fail to exhaust the depths of the other objects with which they share a relationship. Let us focus more closely on an example from above, as related to *Nature*. When a tree is encountered by humans or insects, by the moss growing and spreading out at its base or rain falling from the sky, the tree displays a multiplicity of perspectives to all the objects encountering it, approaching it from differing angles, from unique perspectives regarding *how* the tree might be of concern, "by the insects that swarm to chew on it, by the drops of water sucked into its trunk," and yet the "tree itself is not reducible to any of these properties. And neither is it reducible to the *sum total* of them."[65] Importantly, for our purposes, and to return to a theme from above regarding the *excessiveness* of all objects, for as an authentic object as defined by Harman, in this example, *Nature* is neither reducible to its instrumental *use-value*, i.e., the manner in which it nourishes certain species of insects or might one day serve as fire-wood to warm a woodland cabin on a frigid December evening, nor is it reducible to its bare *physical nature* that is comprised of a multiplicity of smaller component elements. When we consider the world of *Nature*, Harman recognizes that there are "features of the Mediterranean Sea that no currently existing fish is equipped to measure,"[66] no feature or quality that it could ever exhaust, for *Nature* as object shelters its *essence*, forever holds back a *remainder* from full disclosure or revelation. As is also revealed in our analysis, we must keep in mind that the qualities of a tree or mountain are radically different for each entity encountering it, and so too are the qualities of those things or entities that the tree or mountain itself "experiences," e.g., to a mountain-climber attempting to scale the mountain's summit, trudging over the layer of freshly fallen snow blanketing and covering her pathway, which will preserve the imprint of her boots, the totality of the mountain and its qualities can never be fully experienced, known, or exhausted by the climber, and this also applies, in line with Harman's OOO, to the relationship or interaction between the mountain and the snow – and *vice versa*.

To conclude this section we reiterate that *Nature*, as conceived through our reading of OOO, is a real object (RO) with real qualities (RQ), and also a sensual object (SO) with sensuous qualities (SQ), which, it is possible to state, much like all objects, demonstrates the following characteristics: An object "has a retroactive effect on its parts" and generates new parts, and has "emergent properties not found in its components."[67] Yet, we must be cautious here and stress that these characteristics do not exhaust the object by "explaining" it, for Harman assures us that these are just "many symptoms that [inform us] an object is present and these are not the *sine qua non* for objecthood,"[68] i.e., characteristics and properties neither exhaust nor express *in toto* the object's *essence*. Let us briefly take each of these characteristics of objects in turn – and we know that *Nature* is an object – in order to concretize them through basic examples of each prior to analyzing them in a slightly more complex manner: First, related to an object's *retroactive effects*, Global Warming – or "warming-event" – is highlighted by the increase in temperatures because it is traceable to a *retroactive* event of extra greenhouse gasses that are trapping heat at the earth's surface. This in turn is traceable, based on at least 97 percent consensus among recognized, legitimate climate scientists, to the abundance of greenhouse gasses such as carbon dioxide that humans have pumped into the atmosphere since the Industrial Revolution, which is also an object in its own right. Second, as related to objects *generating new parts*, we might consider the issue of how the over-prescribing and over-use of antibiotics to treat a wide variety of infections opens the potential to give rise to new strains of bacteria that possess antibiotic-resistant properties, e.g., the bacteria Staphylococcus aureus is now resistant to most forms of penicillin because this strain has mutated and has developed resistance genes. Finally, with respect to *emergent properties* of objects, we might consider bleach and ammonia, which apart from each other do not possess the same properties or enact the same effects as when they are combined, giving rise to – in terms of an "object-event" or new RO – the highly toxic chlorine gas.

Now, let us return to the EVD with our newfound understanding of *vicarious causation*, along with our initial definition of an object, which includes the above understanding of *retroactive effects*, *generation of new parts*, and the issue of *emergent properties*, as we approach it in terms of an "object-event" that is the result of *vicarious causation* and is still in the process of *disintegrating* and, like instances of causation that Harman considers, demonstrates a sense of *asymmetry* in at least two senses: (1) The difference or gap between ROs and SOs in terms of causation and (2) Encounters between two objects are possible that yield no significant or even inconsequential effects on one of the objects in the event, i.e., one of the objects may have little or no retroactive effect on the other.

Retroactive effects: Initially, the Exxon Valdez tanker sails along the Alaskan coastline because of the manufactured public need and dependence on fossil fuels – indeed, the fossil fuel industry is in and of itself an object demonstrating the traits and characteristics of objects we are discussing – and so the tanker's *presence* serves as one *retroactive effect* contributing to the potential for the disaster to occur in a specific region (*space*), Prince William Sound, and at a specific *time* and date, March 24, 1989.

Generation of new parts: After the occurrence of the disaster, the massive clean-up efforts generated "new parts" of the object-event, which included attempts to save the living animal population affected by the oil spill. And, after the passage of time, the *generation of new parts* includes both the after-effects of the oil spill on the health and recovery of the ocean, beaches, and inland forest areas as well as the ecological legislation proposed and passed in the attempt to prevent such disasters from occurring in the future. One such EPA-sponsored piece of legislation is the Oil Pollution Act of 1990, which includes the clause prohibiting ships that have previously spilled oil in excess of one million gallons from traveling through the Prince William Sound.

Emergent properties: The crude oil within the earth does not hold the same potential for destruction to the environment as when it is extracted, burned, and transported across the land and waterways, i.e., *only* when the oil was transported for sale and consumption and *only* when the tanker carrying the oil collided with a reef along the Alaskan coastline did the object-event's *emergent* potential or property for disrupting and destroying the local ecology arise, come to presence *as* possibility.

With respect to the EVD, there was undoubtedly more significant and lasting damage inflicted upon aquatic and terrestrial plants, indigenous fish and mammals, and waterfowl – to *Nature* – than to human beings, if we overlook or discount the monetary loss that the commercial fishing industry incurred in Alaska, along with the lost profits of the Exxon Corporation. Despite the *asymmetry* present to *vicarious causation*, however, the analysis of this "event-object" demonstrates that in this instance, humans had an observable and pronounced negative influence on *Nature*.

A reconceived encounter with Nature-as-object in "causation" and "aesthetics"

This unique view of objects embraced by OOO initiates us into a view of things, both animate and inanimate, that stand beyond the human being – indeed outside of the *correlation* between thought and world at the center of Western philosophy and ontology – which are not transcendent in terms of a *transcendental signifier* such as God or Being, but rather external, and in an important manner, existing in their own right beyond human thought and determination; we neither control nor master *Nature* as an object in thought or action. This understanding has important ramifications for overcoming the radically destructive and excessively egocentric anthropomorphism that has already been identified as giving form and structure to the contemporary CF, which, as we have seen, attunes us to an incredibly shallow and impoverished view of both objects and *Nature* as a vast collection of things, objects, and *resources* that are manipulable, useable, and disposable. This alone should call us to question and reconsider the manner in which we respond to *Nature*, the manner in which we assume a stance before its awesome force and power, inspiring the re-conceptualization of the *correlation* that is human intentionality, for our ultimate concern is with eco-phenomenology and the human-and-*Nature* relationship. This is precisely the line of inquiry we pursue when moving through the remaining chapters in the quest to learn *how* we "connect" with *Nature* in terms of experiences that might be classified as ethical, aesthetic, or even *poietic*, which are communicated through *non-literal* language, in a way that for Harman, incorporates the experience of "metaphor," an aesthetic medium through which to appropriately approach the bare essence of *Nature*. Through the immediacy of our *intentional experience*, as Harman observes, we commune with the objects of the world in an aesthetic/philosophical manner, which elicits a powerfully transformative effect upon us, an effect that we argue occurs in the form of an *ek-static* mode of *attunement* that situates us in the presence of *Nature*'s *imperatives* inherent in the *things themselves*, as we find in the phenomenology of Lingis, an experience that is imbued with the undeniable aura of *normativity*.

When Harman outlines the components of intentional objects (IO), he claims that objects stand or relate to each other in one of the following five ways: (1) *Containment*, which indicates that both objects are located within the interior of a third object; (2) *Contiguity*, which indicates the temporal succession of the side-by-side relationship of objects within the experience; (3) *Sincerity* – and this is the component that will be of our concern – which indicates that the experience elicits a mode of *fascination*, or as Harman puts it, that we become absorbed in the sensual objects of our intentional experiences, e.g., "I am fascinated by the sensual tree, even if my attitude toward it is utterly cynical and manipulative"[69] – and it is here that we are interested in relating *sincerity* to the type of *attunement* we spoke of in earlier chapters; (4) *Connection*, which instantiates the creation or birth of a new object formed from the intentional experience; and (5) The *case of no relation at all*, which indicates, as we have already stated, that certain objects simply do not relate at all to other objects. When seeking the most important of the five components related to the root of all change, Harman concludes that *sincerity* is "the site of change in the world,"[70] for when a real object relates to other objects within an IO through the revelation of its *sensual face*, it gains oblique access to the *thing-in-itself*, "piercing their colored mists" and connecting with a "real object already in the vicinity but buffered from direct contact."[71] This is what we have referred to throughout as the phenomenon of an object's *withholding* or *withdrawing* from "knowledge," but offering hints and intimations of its *essence* in a concealed and incomplete way through *allusion*; indeed philosophy itself is "the great art of allusion to the real that cannot manifest itself directly."[72] In order to explore *sincerity* in its intimate relation with the phenomenon of the *allure of objects* and the *effect* that their *volcanic core* has on other objects, which include human beings, we consider Harman's groundbreaking work on philosophy *as* aesthetics and metaphor as a descriptive medium for bringing us as close as is possible to the naked experience of an object's *essence*.

Harman refers to the interior of objects as a "volcanic core," and in addition, when referencing the inaccessible *essence* of objects, he employs such visceral and emotive terms as "kaleidoscope, witch's cauldron, steel mill, or alchemist's flask in which one thing is converted into another."[73] It is the interior of a new object created by the relationship between two ROs that is the *primordial* "site of reality,"[74] and this *reality* is precisely what goes unnoticed in acts of knowing that we have associated with *undermining*, *overmining*, and *duomining*. What we explore now in relation to intentionality and the intentional object (IO) will focus on the correlation between human and world or *Nature* – this considering our project is concerned primarily with eco-phenomenology – with the important understanding stressed throughout that relationships between objects exist beyond the correlation between thought-and-world. It is possible to understand intentionality as an *object-creating* phenomenon in the following manner: My intention of another object (tree, animal, human, or environmental disaster) is an encounter with another real object (RO) and its sensual counterpart, and the relationship unfolds "in one place only; the interior of another, larger object,"[75] i.e., the intentional relation as a whole – as new RO – *is* the encounter between two sensual objects (SOs) and their sensual qualities (SQs) in tension with their respective real objects (RO). When two objects meet, the experience consists in two sets of SQs "pressed against each other, encountering one another in more or less turbulent fashion,"[76] which might be traced back to the SQs of the SOs given form by the RO, which always withdraws from direct contact with the other RO. In addition, the RO also withdraws from the SQs of the other object, and it is the sensual qualities (SQs) of both objects, intermediaries presenting their "sensual faces" to each other, that actually meet in the encounter. This *sensual face*, as we have named it, serves as the *intentional object* within the encounter, but there is now another object, which is created and instantiates what we have already called a *real object*, and this, to return to what we stated above, is the unified intentional act that contains and shelters *human-and-tree* within its interior, i.e., both objects "reside in the inner molten core of a [newly produced] *real* object,"[77] and like every other RO, "it is a reality whose full depths can never be exhaustively probed."[78] In this intentional act, I am in the presence of the RO's (e.g., the tree's) *essence* radiating out toward me, and indeed seducing me into this encounter, and this is because the object, "beyond all its *essential* qualities, is more like a brooding power or style that lurks beneath the qualities and animates them."[79] Thus, the *interior* of the new object – the intentional act as a whole, the encounter and relationship between human-and-tree – is a "perceptual space filled with intentional objects rather than real ones," and importantly, this interior space is active, and like a dynamic vortex, it holds and preserves "other relations that might one day be generated."[80]

In relation to *sincerity*, this intentional encounter as described exacts an effect on me, by inspiring a sense of *fascination* that can be related to what Harman calls "allure," because the experience is of a powerfully inviting and captivating nature, which is irreducible to either *theory* or *praxis*. The experience of *sincerity* occurs as an object is given in its most primordial manifestation, as that which resists the two poles of explanation commonly employed to exhaust an object in "knowledge," defying reduction to either its cluster of qualities or its use value, inhabiting a *third space*, as introduced at the outset of the chapter. And in this encounter, which we must keep in mind is an instance of *causation*, both objects are given over to a phenomenon that is beyond them, the "object-event," and thus both are "built out of something even more primitive,"[81] and when we enter or are drawn into this *interior space of the primitive*, we are attuned. So, here we return to an idea from Chapter 1 based on Harman's reading of Lingis's phenomenology. Revisiting the essence of the relevant passage we originally quoted, let us consider *how* it might relate to our current discussion of *causation* and *sincerity*: When out on a hike and approaching the base of a steep mountain in the wilderness, our immediate and intuitive perceptual experience is irreducible to the common, everyday understanding of *two natures*, i.e., irreducible to our encounter with the mountain's physical exterior presenting as a towering mass of gray and black layered matter, or to an understanding of the mountain emerging from knowing its interior organization, the mountain-specific atoms, the specific minerals deposited on its surface and penetrating into its depths, or the precise ratio of organic and inorganic matter composing its massive rising structure. Rather, Harman asserts that we experience the raw "style" of the mountain, the "effect" of the mountain, i.e., we are moved and attuned by the effect that it has on us, as we stand awestruck, swept up into the swelling presence of its undeniable "grandeur." Within the "object-event" instantiating the encounter between the mountain and me – the new *intentional object* sheltering both "objects" within the interior core – I am transformed because of this *effect*, and in the following chapters, we develop this experience in terms of our *coming-into-the-presence* of and standing-within *Nature*'s integrity, which is ineluctably bound up with our *attunement* to the intrinsic value of *Nature*, which we argue can be ignored or defiled by the human, but never given or taken away. It is crucial to note that the mountain, in this experience, has also been *transformed*, for within the attuned intentional act it has been released from the strictures of our common, everyday, and *naturalistic* interpretations and understandings, and so the experience holds, as we explore later, not only possibilities of an *aesthetic* nature to manifest, but also the potential for such attuned moments to break open a space for the *ethical* to emerge.

We now focus on the importance of aesthetics in Harman's OOO, as this will further contribute to our understanding of *sincerity* and *allure* in its relation to an "aesthetic effect." We have already stated that we know objects – and objects "know" other objects – only by means of *allusion*, or through oblique intimation, in that we never experience or "know" an object's *essence*, its *molten volcanic core*. The effect of *alluding* to or *gesturing* toward an object's *essence*, and to actively participate in this *allusion* through a descriptive medium, is to experience the primal *allure* of objects that is unavailable through other means – this is also expressive of an experience, as we detail below, that Harman calls "*sincerity*." Harman links *allure* with an aesthetic effect, and for OOO, in a legitimate manner, *aesthetics* can serve as "first philosophy."[82] In and through aesthetics the possibility exists to *approach* the *essence* of objects without replacing the experience with an attempt to "know" objects employing literal language. We do not associate aesthetics with the realm of "knowledge" proper, and indeed, for Harman, philosophy is closer to art than it is to science, for "philosophy and the arts are forms of cognition without being forms of knowledge,"[83] and art gives us what Harman terms, "justified untrue belief." In the next chapter we locate the place of art and ethics within *axiology*, along with the mode of expression (argumentation) common to both. It is through the aesthetic experience that the inward nature of things is opened to us, but always in a manner that defies literal explanation, in a moment we equate with a *veiled shining forth* of *essence* or the *thing-in-itself* of objects. J. W. V. Goethe, in relation to the issue of Nature's propensity to hide its secrets, expresses this metaphorically and poetically: "Nature does not suffer her veil to be taken from her," and in her recalcitrance, in her choice not to "reveal her spirit," neither by "levers" nor "screws" shall her essence be wrested from concealment.[84] For OOO, aesthetics deals with "ghostly objects withdrawing from all human and inhuman access, accessible only by allusion and seducing us by means of allure,"[85] bringing us face to face with the object's primal power that exceeds any of its properties, and in and through the aesthetic medium, those properties are *animated* from within, as "some ill-defined demonic energy" is conjured up and set loose, released to permeate the aura of the experience.[86] Harman contends that *allure* "splits an object from its qualities," and it is primarily through the experience of metaphor that this phenomenon occurs in a most powerful manner. In the aesthetic experience of metaphor, "the debris is cleared away and the object is encountered as flickering at a distance, in all of its naked allure."[87] In metaphor, it is possible to set objects free – in and through the aesthetic process of *standing in* for them – as we shall see – releasing them from the grip of literal explanations, and thus from reductive forms of knowing associated with *undermining*, *overmining*, and *duomining*. But how does metaphor accomplish this? In responding to this query, an account that closes this section, we analyze Harman's unique understanding of metaphor's crucial role in philosophizing the ontology of objects.

Two of Harman's favorite metaphors are derived from Homer: "The wine dark sea" and "dawn, with her fingertips of rose." However, we might also turn to the poet and philosopher Friedrich Hölderlin, whose poetry compares the "sunset" to an "evening hymn" played on the "heavenly lyre" of the sun god, *Helios*.[88] Indeed, as related directly to our talk of the EVD, we find the following metaphor from Tony Ogunlowo's poem, "When Oil Spills," which reads: "Like the Rivers Of Hades/ the black sludge spreads forth/ consuming all in its path." Since, as Harman claims that neither *theory* nor *praxis* provides a clear, non-reductive mode of seeing into the authentic essence of things, even our pre-reflective immersion in the world to which phenomenology returns us is also a reductive and dissembling mode of world-disclosure. When we describe things and participate in poetic depictions – as spectators-actors – which are metaphorical in nature, both the things described and we who are drawn into the experience of metaphor are transformed. For Harman, as stated above, it is the aesthetic experience that sets us within a close ontological proximity to the essence of objects, which establishes an intimate and transformative type of intentional relationship where we experience and are moved by the *allure* of the world. In moments of aesthetic experience or aesthetic attunement, as we have used this term, the world becomes *more real* in an undeniably visceral way. In relation to the quadruplicity associated with Harman's *flat ontology* of objects, he argues, "metaphor can be analyzed as an RO-SQ pair,"[89] or what we have termed the tension between a *real object* and its *sensual qualities* that emanate from it and are traceable to it. With the aesthetic experience, these qualities are rendered in a new and transfigured light, an experience that metaphor inspires, and cannot, in order for the metaphor to work and hold its seductive power to sway and entice us, be traceable in a literal way back to the SO in connection to the RQ and RO. This, for Harman, "holds true for all aesthetics, which differs from normal experience," e.g., as in the description of *vicarious causation* from above, "in presenting a non-literal object rather than a literal one."[90]

To understand how metaphor functions along with the subsequent attunement that accompanies the human's aesthetic experience of the world, which is the experience of creating a new "aesthetic" object and "not knowledge about pre-existing objects,"[91] we must consider that the *intentional aesthetic relationship* metaphor instantiates is consistent with what we will term the enlightened, "creative-making" associated with *poiesis*. When considering Homer's metaphor "dawn and her rosy-fingertips," dawn assumes the role of a RO and her fingers assume the role of SQs, but they are linked to a SO in a way where there is a "collision between a familiar object [SO-RO] and unfamiliar qualities," which cannot easily be reconciled and which throws any and all literal readings into confusion, for the "finger-tipped dawn is a mysterious one in a way that an everyday dawn is not."[92] In this aesthetic relationship between SO (as linked to the RO) and SQ, dawn as RO is aesthetically reconceived in terms of a "personified metaphor," and it "stands at a respectful distance from us, commanding our interest, withdrawing into shadow."[93] It is also the case that in metaphor a new group of SQs are attached to an SO that is related to the RO, but because the SO cannot uphold – because it is inconsistent with – such non-literal qualities, it disappears from the relationship, and so "a mysterious real object is needed to do the job,"[94] i.e., to fill in and support the metaphorical sensual qualities. However, since we know that ROs *withdraw* from our grasp, we are left in the case of metaphor with what Harman identifies as *free-floating* qualities in the absence of an RO. Harman's radical claim is that in metaphor, "the sensual qualities of the metaphor are supported instead by the only RO that is not withdrawn,"[95] and it is the case that in metaphor the human being steps in and assumes or *plays* the role of the RO, in terms of a participant-beholder, i.e., in a *theatrical sense*, the human becomes a *thespian*, an *actor*. We are drawn into this relationship and it is the metaphor that facilitates this occurrence, for we become the aesthetic "stand-ins" replacing the RO to which the metaphorical SQs are now attached. For Harman illuminates the phenomenon of metaphor in terms of an intentional relationship between human (RO) and aesthetic SQs, and when considering Homer's metaphor, we are at once the *readers* and the *method actors* "playing the dawn playing fingertips."[96]

Through metaphor we experience and participate in the *poietic* creation of a new object, and this is, as we now understand, also an instance of *vicarious causation*. In and through our encounter with metaphor, as Harman contends, "metaphysics may be a branch of aesthetics, and causation merely a form of beauty."[97] Aesthetic experiences bring us in touch with the world in a way that lies beyond the epistemic bounds of either everyday or technical and literal knowledge of the world we inhabit, bringing us closer to things in a mode of "touching without touching,"[98] and importantly, this *indirect* access occurs by way of allusion. In such moments we are separated off from our everyday involvement with objects in causation, for art can "give us something like the thing in its own right: the infamous thing-in-itself."[99] When seduced and moved by the *allure* of things, we feel, perceive, and experience things differently, in a non-literal manner, in a way that facilitates the shining and revealing of things in a different light – *allure* instantiates a transformative event of *attunement*. Harman is emphatic that in the experience of Homer's metaphor there is no epistemic "justification" for believing that dawn ever really reaches upward with her outstretched rosy-fingertips, but because we reside and dwell within the interior of a new aesthetic object, we revel in its descriptive and imaginative beauty and it becomes an aesthetic possibility and we are awestruck. This also occurs when we participate in the poem by Hölderlin: We imagine that the sunset we are experiencing is like a song produced by the holy lyre of *Helios*, and we are overtaken by the harmonious tones that the god produces. And, finally, when reading Ogunlowo, we might experience the aesthetically transformed "horror" of an oil spill, when millions of gallons of crude oil flood into the sea from a large gash torn in the side of a tanker, much *like* the rivers of Hades, engulfing and destroying the surrounding environment as a primordial and unholy force is unleashed. Harman is clear that the aesthetic experience is not to be understood in terms of fanciful or playful excursions within an active imagination, because we *whole-heartedly believe the things we encounter and feel in metaphor*. For as he reasons, in aesthetics we cannot have "truth" proper, but we are certainly justified in believing and trusting our experiences, this because metaphor "*creates* its own object in the very act of *believing*."[100] This is why Harman refers to art as offering "justified untrue belief," and in Chapters 4 and 5 we return to this notion in its relationship to *ethical intuitionism*, which we claim is inseparable from, because it is inherent in, our ethical-*normative* experience of *Nature*.

Enlightened return to issues of eco-phenomenology

In this final section, we consider what might be learned from Harman's OOO in relation to our developing eco-phenomenological project, with the understanding that we are not adopting Harman's ontology *in toto*. Yet, this is indeed how authentic philosophy attempts to find its way as it moves through various permutations in new and unprecedented directions, drawing from its history with a critical eye to its reinterpreted future, e.g., Harman draws heavily on Jose Ortega y Gasset's analysis of metaphor when formulating OOO's original reinterpretation of that view. Harman also draws from the theological tradition of religious *occasionalism* found in the Islamic thought of Abū al-Ḥasan al-Ash'arī and al-Ghazali, and convincingly offers a unique version of philosophical realism that includes the secular transformation of these theological ideas in order to philosophize OOO's notion of *vicarious causation* as presented. At a fundamental level, Harman agrees with our position regarding the oppressive *presence* and *attunement* of the dominance of the *metaphysics of presence* (CF). Harman's unique understanding of the modes of "knowing" expressive of *undermining*, *overmining*, and *duomining* is a reminder that we must always be skeptical and wary of all modes of human knowledge, and especially modes of knowing that stubbornly and dogmatically embrace the conceptual lens (CF) grounded in the epistemic mode of *positivism*; for us, recall, this tendency is evidenced in *secular humanism* and the erroneous belief in unadulterated *human progress* through the unimpeded growth of *science and technology*. This skepticism is crucial for any renewed vision of *Nature*, e.g., we agree with Harman that an understanding of objects and their relations, along with our interaction with those objects, can never be reduced to *materialism*, which holds fast to the false belief that all things in the world *must be physical*. Indeed, the notion of "matter as unformed physical stuff in which forms can be stamped has no basis in experience, and is simply unhelpful in theoretical terms,"[101] and as related to eco-phenomenology, this disingenuous view leaves no place for the authentic consideration of ethics in our dealings with objects, things, and *Nature*. The following four issues from Harman's ontology contribute to furthering our project: (1) OOO's radical move beyond anthropocentricism; (2) Harman's suggestion for incorporating non-traditional sources for inspiring rich textual analyses in relation to a developing philosophical project; (3) The manner in which the notion of *justified untrue belief* in aesthetics is transposable to the realm of ethics – *axiology* – and, (4) The crucial role that *attunement* plays in transforming our mood, thought, and action.

As stressed throughout, eco-phenomenology seeks to transcend both the understanding of *Nature* as a conglomeration of physical masses dancing to predictable laws and the concomitant view that we can master these laws by extracting *in knowledge* their essential form, and that in doing so, we hold the supreme power to intervene, to control and manipulate *Nature* to our anthropocentric demands and needs. Harman outstrips even the most entrenched, pervasive, and ardent forms of anthropocentricism, for OOO radically de-centers the human being in a manner that is more extreme than the reconceived notion of phenomenological selfhood emerging from eco-phenomenology. It is accurate to state that in OOO humankind is dethroned, it is recast and reinterpreted in a context where it is on equal footing with the entirety of *things of the world*, it is no longer *king of the universe*, and this we trace to Harman's acceptance of *flat ontology*. This sets OOO against what we might term "modern" ontology as is consistent with the CF examined earlier, for Harman treats all things equally and is resistant to acquiescing to "personal prejudices about what is real or not," for an authentic philosophy should be inspired to explore "features that belong to *everything*."[102] For example, according to Harman, modern philosophy expressive of traditional metaphysics embraces a "shoddy dualism" that includes humans (as privileged in the equation) on one side and the entirety of all other things in the world on the other. Harman insists that we should avoid the typical metaphysical move to "automatically make human beings worthy of filling up fifty percent of ontology."[103] Although Harman *looks outside* of the traditional correlation between human thought and world to consider the world of objects that exists outside and hence beyond it, he is not suggesting we can *truly* or somehow authentically "think" outside it – OOO is obviously *not* suggesting some preposterous New Age view of the disembodied conscious mind – but what he does is to inspire our thinking in a way that enhances and contributes to the re-conceptualization of our traditional view of the human's place in the world. For after moving through his radical ontology and analysis of objects, we return in an enlightened and transformed manner to productively, and we argue, *ethically*, rethinking our traditional views about the issue of *correlationism*. For example, conscious intentionality, or *immersed-intentionality* as we have called it, can be understood in terms of a specific way of inhabiting *space*, in terms of what we have identified as ontological *spatiality*, or the proximity to *essence*. Based on our detailed reading, the relationship between human-and-world is an intentional relationship that is an "object," a *third object*, in its own right, with boundaries that are formed and sustained only in connection to the so-called "outside" world of objects, which is formed through the connections and relations between objects. This issue will be thoroughly explored in the final chapter when we explore *immersed-intentionality* and the phenomenology of the human-and-*Nature* relationship.

Simply embracing the idea that there are things and events existing and occurring in the universe that are beyond the human's control, beyond its thought and volition, while appearing inconsequential at first blush, is enough to get us thinking in a direction away from anthropocentricism, in a manner that can be extremely humbling and potentially beneficial for inspiring an anti-anthropocentric stance, contributing to a renewed vision of the intentional relationship that the human-and-*Nature* correlate instantiates. Importantly, Harman stresses that the idea of a legitimate world and universe – a *reality* – that exists beyond us is foreign to *metaphysical anti-realism*, and as Harman has shown, there are many interactions in the world, instances of *vicarious causation*, occurring with regularity that do not in any way involve or affect humans. *Nature*, as we understand it, is an *object* that instantiates a significant and powerful mind-independent reality that *antedates* us, precedes us by millennia, and will surely *postdate* us; it will succeed us and continue to exist long after our species goes extinct, long after we have spiraled off of this precarious, ephemeral, and fragile mortal coil. Such a view related to our quest for an eco-phenomenological ethics enhances our thinking on *selfhood* and finite *transcendence*, for when *becoming other to ourselves in the face of the other*, and hence *transcending* our present incarnation, or mode of Being-in-the-world, we are recognizing, embracing, and demonstrating a solicitous care regarding the *good* or *value* of something beyond ourselves, which we deem necessary for living a "good" life in concert with *Nature*.

It is from Harman's extraordinary work on metaphor that it is possible to draw inspiration to enhance the descriptive interpretations of eco-phenomenology; this is because Harman expands the way we typically encounter writing in philosophy. For as opposed to rigorous and literal treatise, Harman challenges us to consider the "philosophical" value that might be found in many forms of writing as potential inroads for acquiring a deeper understanding of the world, or in our case, *Nature*. Recall that in our discussion regarding the transformative power of metaphor, as is consistent with aesthetic experience, several short passages from Homer were analyzed and we came to see that through our immersion and participation in metaphor a new "aesthetic-object" is created in which we are intimately involved, and this experience of being on the *interior* of a new object, in a moment of *attunement*, which for Harman elicits the mood or feeling of *sincerity*, transforms the way we experience and inhabit the world, e.g., our new "aesthetic" experience with a mountain momentarily transcends any literal or formable knowledge of it. Harman shows that we learn something radically new about the natural world in a way that transforms us from Homer's poetry through the experience of metaphor! Harman's work on metaphor is rife with potential for contributing to and enhancing the practice of phenomenological description and analysis. For example, as introduced in Chapter 1, in eco-phenomenology we typically encounter either an informal method of analysis or a more complex practice of the method, which includes the *noesis-noema* correlate, beginning with the "shift from the natural to the phenomenological attitude," employing the *epoche* (bracketing) for "bracketing the bounds of familiarity," after which we perform the process of *eidetic variation* in order to arrive at a "new level of familiarity" or "level of essential seeing" in terms of the *eidetic reduction*.[104] However, these two methods do not exhaust the ways we gain access to the pre-reflective realm of existence that eco-phenomenology seeks, and Harman's work on metaphor suggests that it is possible to consult literature, poetry, and other viable literary and philosophical sources, which offer imaginative descriptions of "lived experience," in search of new and unique instances of "lived experience" for our eco-phenomenological analyses. This is an issue that is often ignored in both phenomenological and eco-phenomenological studies, for it is the standard practice for the author, philosopher, or scholar to provide her own "pre-reflective" and "subjective" experiences through description for analysis, and while this practice undoubtedly yields rich and valuable studies, if limited *only* to our own "first-person" view and experience, other opportunities for enhancing the depth and appeal of the work are foreclosed.

Here, in relation to Harman, we introduce several phenomenologists who follow this track of consulting *alternative written sources* for philosophical inspiration. Perhaps the most famous phenomenologist pursuing this line described is Heidegger, during and after the 1930s, once he graduated beyond the *fundamental ontology* of Dasein in *Being and Time* and his strict incorporation of method and dense, technical academic jargon. In relation to this issue, George Pattison asserts that, "increasingly in his middle and later work, Heidegger is less and less concerned with the kind of questions concerning perception that occupied his early phenomenological endeavors," and far more interested in exploring the issue of "what gives itself in language and, above all, in texts."[105] We must note that although "Later" Heidegger eschewed a technical phenomenological method, he nevertheless continued to practice what might be termed *phenomenological-ontology*; he just referred to it as "thinking," which primarily, as Pattison emphasizes, revolved around the detailed and critical reading of texts, both philosophical and poetic. Phenomenologist van Manen also expands his use of "phenomenological sources" to include, in order to supplement and enhance descriptions of direct first-person experience, "literature, biography and other artistic sources," which "may provide us with powerful examples of vicarious lived experiences and insights normally out of the scope of our personal everyday experiences."[106] Based on our concerns, this practice opens the possibility of gaining legitimate phenomenological insight into *Nature* when consulting such sources as poetry, novels, stories, and plays, which opens phenomenology to embrace aesthetics as a source for lived experiences. It is also the case that we might consult interpretations of "lived" experience found in *existential philosophy*, e.g., Sartre's descriptions of *love*, *absence*, *shame*, and *freedom* in *Being and Nothingness*. This is one crucial reason for engaging Lingis's rich, visceral, and unique poetic descriptions and phenomenological interpretations, and this is an instance, for van Manen, where the work of "*other phenomenologists turns into a source for us with which to dialogue*," offering us the unique opportunity "*to reflect more deeply on the way we tend to make interpretive sense of lived experience*."[107] Directly related to this issue in eco-phenomenology, Brown gestures toward the writings of authors and poets that might legitimately provide us with instances of moving descriptive material for analysis, by suggesting *non-traditional* phenomenological sources such as "literature," which he embraces as a fecund medium for enhancing an understanding of ecological-ethics, empowering us to break free from traditional anthropocentric discourse, which in addition to its typical *value-free* rationality, in relation to Harman, retains the tendency to embrace and communicate through literal and formal modes of discourse. As Brown insists, we can learn and develop by reading

> a great deal of nature writing, from Thoreau and Muir through Leopold and on to today's Radical Ecologists, attempting to establish a new mode of moral and aesthetic discourse in which the experiences of the intrinsic goodness of nature can be registered, expressed, and rationally developed. Without such a vehicle of articulation, experiences remain mute and powerless and are dismissed to the margins of rationality.[108]

Harman's philosophy makes an important contribution to Continental philosophy, aesthetics, and architecture, with his unique view of *philosophy-as-aesthetics* and *aesthetics as first philosophy*, and Harman recognizes that much of the "credit for importing OOO into architecture goes to David Row, now of the Southern California Institute of Architecture."[109] It is possible to read Harman's view of aesthetics as challenging, in addition to Kantian formalism, Heidegger's transcendental view of aesthetics and attunement. For although Heidegger associated art with a change to one's Being-in-the-world through *attunement*, he ignores the view that there are numerous instances of great art that do not, because they cannot, rise to the level of cultural or historical grounding and founding forms of art. Drawing from Harman's ontology and view of aesthetics, we can imagine many instances of art, such as plastic art and cinema, that hold the power to transform our understanding in small, but certainly not inconsequential, ways. Reading Harman, we find that forms of art that fall short of Heidegger's stringent and highly reductive classification undoubtedly hold the potential for enlightening us and inspiring moments wherein we are moved to challenge and overcome views in the history of philosophy that have been codified and dogmatically ossified. Art need not found and ground historical cultures to be thought of as a viable, valuable, and transformative force that holds the potential to enrich and contribute to enhancing and changing our lives. As stated, OOO claims that art can open us through a form of *attunement* (*sincerity* and *allure*) – through *indirect* and non-literal access – and set us in close proximity to the elusive and mysterious *thing-in-itself*, a view that transcends both the philosophical aesthetic tradition and the modern aesthetic criticism or theoretical tradition. Although Harman does not make this claim, we might legitimately infer that OOO also opposes what might be classified as a strict *aesthetic cognitivism* as found in Nelson Goodman, who situates aesthetics under the philosophical auspices of *epistemology*, classifying art as a legitimate form of knowledge. As we have seen, Harman denies that art produces anything relating to a traditional understanding of either "knowledge" or "truth," and this, turning to Terry Barrett, whose analysis of aesthetics is in line with that of Harman, demonstrates the potential of "aesthetics" to "offset the positivist (realist) tendency to dismiss emotive expression as meaningless."[110] Although the following comparison cannot be developed in this short section, the view of aesthetics that OOO philosophizes boasts several key elements contained in the view of art of philosopher Suzanne Langer, who would be vehemently opposed to the view of metaphor held by the aesthetician Mark Johnson, who argues in favor of realism in art and meaning, and in the process heedlessly dismisses metaphor as useless in inspiring and facilitating the *experience of art*. As we now know, contrary to Johnson, whereas in phenomenology it is description and analysis that brings closest the primordial *essence* of the things we encounter in our pre-reflective experiences, for Harman, our movement into the ontological proximity to the *essence* of things occurs most powerfully and profoundly through our active participation in the aesthetic medium of metaphor.

Within the aesthetic experience, we are immersed and overtaken by the encounter with a transfigured world, in one important way for Harman, through our use of and participation in metaphor, and in this experience: (1) We are brought in close proximity to the *thing-in-itself*, and, as stated, Harman claims this happens when we find ourselves on the interior of a new "aesthetic object" –this we now know as the *volcanic core* of *reality* and the experience itself is monumentally transformative; and (2) Because art is a form of non-knowledge, or *justified untrue-belief*, our experience stands beyond reductive forms of knowing the world, i.e., beyond the distortion and dissembling of *undermining* and *overmining*, and their subsequent unification in *duomining*. However, according to Harman, art inspires a legitimate form of cognitive activity, but is not reducible to a discursive kind of cognition, and beyond this, "it is a form of cognition without being a form of knowledge."[111] Here, let us recall the phenomenological analysis of the "tree effect," which effectively represents a moment of transformative attunement and *insight* that Harman highlighted in his reading of Lingis's phenomenology as introduced briefly in Chapter 1. When speaking of *justified untrue-belief*, we must note the following: This mode of *insight* into things stands at a distance from what Harman identifies as *justified true-belief*, which is related to the attempt to know a thing's composition and its function or use, and further, knowledge or truth that claims "direct access to the real."[112] The problem Harman identifies with *justified true-belief*, and here we must keep in mind the function of propositions, is that it relies on an impossible correlation "between our beliefs and reality," a *correspondence* that can never be established with the type of certainty to guarantee "truth" as a condition for *having* and *claiming* knowledge. Based on Harman's analysis of the radical epistemic shortcomings of *undermining*, *overmining*, and *duomining*, truth is not possible in the sense that is traditionally understood and accepted in the history of Western philosophy, which is why Harman makes the radical claim that *truth is impossible*. When examining Harman's notion of *justified untrue-belief* we learn that it is not as outrageously controversial as it appears when contrasted with the traditional model of truth, for we speak of what is "untrue" in the sense of "knowing" – we prefer "intuiting" – in a way that recognizes "we are dealing with a reality that eludes us," rather than giving itself to us, or revealing itself, in terms of "a directly accessible truth," i.e., it is a "knowing" that fails to "mirror the world exactly."[113] We can say that this form of "knowing" is *justified*, however, because "its unstated principles" hold enough veracity to "empower a great deal of fruitful work,"[114] in other words, as is evident from the many and varied life-projects we pursue, that we can rely on and successfully build upon the insight that this form of aesthetic "intuition" provides. Indeed, as we will see in Chapter 4, *justified untrue-belief* is related not only to the realm of aesthetics, but more importantly for our purposes, also to the realm of the ethical. We develop this idea in relation to the *axiological* attempt to provide "proof" or verification for our moral claims, for when dealing with ethical matters, we rely on *normative statements* when making claims and formulating arguments, which cannot be equated with propositional statements of "fact," or what Harman classifies as *justified true-belief*, although, as Harman recognizes, these arguments are still *powerful enough to legitimately inspire a great deal of fruitful work*. It is to our analysis of *axiology* and its unique philosophical focus that Harman's OOO contributes as we approach and analyze the potential of environmental ethics through eco-phenomenology.

We conclude with a salient ontological theme drawn from Harman, the issue of *attunement* – as a mode/mood of transformative *attitude* – which is intimately bound up in the aesthetic experience of metaphor as concerns our eco-phenomenological project, in light of the weight we place on the exigent need to transcend the insidious, ever-deepening attuning grip of the CF. To begin, recall Harman's claim from Chapter 1 that, for Lingis, "even inanimate objects demand that we treat them in some particular appropriate way,"[115] and for Harman, this crucially indicates, although this line of thought is not pursued or developed, that the *appropriate way* to treat things is inseparable from a moral attitude, as it is expressive of what is *ethically right* or *wrong*, and this is an issue we take up and develop fully in Chapters 4 and 5. We note that Lingis's view runs counter to the idea encountered in Kant's deontological Moral Law, namely, that it is only humans that should be treated ethically and afforded an unconditional sense of dignity and respect; for Kant, "there is no ethical dimension at all to how we treat mere things."[116] Since, as stated, we move to address the issue of the *ethical imperatives* encountered in *Nature* and the manner in which these manifest, this marks the continuation and expansion of Harman's argumentation within his reading of Lingis's phenomenology that originated in a short, but extremely valuable essay, "Alphonso Lingis on the Imperatives in the Things" (1997), which can be found in the collection of Harman's early essays cited throughout these chapters, *Towards Speculative Realism*. Recall that Harman observes, "Lingis takes a step that never occurred to Levinas: the structure of the imperative, it is claimed, lies even in the things themselves,"[117] and this indicates that the things we encounter, through *immersed-intentional* acts, split into two divisions, first, an interior "motor schema," and second, an outward aspect, showing a radical division that "no longer belongs to the human being alone."[118] Harman, we must add, *explicates* in his later fully developed philosophy of objects (OOO), as we have seen, *a step that is boldly taken by Lingis, but never developed into a fully formulated or articulated position*, namely, that the imperatives in the *things themselves* are not only experienced by humans, but also by the things of the world, i.e., the "imperative structure" is also present to the lives of "wolves and dolphins, to the zebras racing across the savanna and the ravens playing pranks with clotheslines"; indeed, even the more "widely despised organisms" are "governed by an imperative as well."[119]

Let us explore in more detail Lingis's moving passage presented in Chapter 1 as it related to our view of *Nature*, which highlights and reinforces Lingis's skill at presenting beautifully provocative and moving phenomenological descriptions. By referencing them as "moving" we are of course indicating that they awaken and foster a change of *attunement*, which we might say, through its undeniable aesthetic and ethical (*normative*) force, wrests, or better, calls forth from out of *silent concealment*, from out of the background into appearance, what we have referred to in Chapter 1 as *themes* of phenomenological concern or Sense (*Sinn*)-giving-making structures. Recall that our understanding of *essence* in its relation to *theme* is a re-conceptualized view differing from the metaphysical tradition, for example, as we find in Husserl, and in addition, as is evident from our reading of OOO, as we encounter in Harman's understanding of *essence* as specifically related to the tension between objects and their qualities. However, our notion of *theme-essence* is in line with Harman's contention that the ontological phenomenon of *essence* defies all attempts to completely understand it, i.e., a *theme* can never be understood in categorical terms, in terms of *justified true-belief*. As introduced in Chapter 1, the reader now knows that our conception of *ontological themes* is reminiscent of, to point to one prominent and well-known example, what the German philosopher of *Existenz* Karl Jaspers referred to as "ultimate situations" (*Grenzsituation*). These are *existential* situations giving structure to the human condition that at once serve as ontological *themes* for our philosophizing. However, as Jaspers emphasizes, we cannot avoid, evade, or change these *themes-situations* and, we must add, they never appear as *problems* or *issues* that can be eradicated, completely understood, or exhausted epistemologically.[120] Rather, we are condemned to bear up these situations, acknowledging them; and the best we can hope for is to gain a deeper understanding of them. *Grenzsituations* manifest in the form of such *ontological truisms* as: *We must die, suffer, encounter chance and misfortune, struggle under the inordinate weight of guilt*, and, we importantly add, in relation to eco-phenomenology, *we are forever bound up with our ineluctable responsibility to Nature in that we are inhabitants of the earth and are therefore dependent on it and beholden to it*.

Lingis's poetic phenomenological description of orchids in the forest inspires our eco-phenomenological contemplation of the notion, because it induces a transformative mode of *attunement* or "effect," as Harman claims, that the natural world is importantly irreducible to a mere mass of resources awaiting our intervention in order to transform it and harness its power for our *instrumental purposes*. Participating in this description of the orchid in the forest, as we are seductively drawn into the passage, we experience the sense of *value* that is already present in *Nature* and that is antecedent to any human involvement, i.e., outside of the valuing subjective mind of the human projecting its values onto *Nature* as a value-neutral phenomenon. This view, which links and traces *Nature*'s value exclusively to the valuing human subject, heedlessly ignores the undeniable sense of *normativity* present to and pervading the intimate forest dwelling, the ontological *space* of the forest. Let us read the following description that is culled from a longer passage: "Orchids are plants with atrophied trunks and limbs, parasitically clinging to the rising trunks that shut out the sun, flowering their huge showy sex organs, awaiting the bees for their orgasmic unions."[121] Harman provides us with one legitimate "phenomenological" interpretation of Lingis, through which he offers a stinging critique of technology viewed as a problem as it is instantiated within the worldview structured by the *metaphysics of presence*, for in this description, Harman claims that Lingis "de-fetishizes the object, displaces its usefulness and manipulability," in a way that "leaves behind the original object as a useless, but gorgeous flower, as an orchid."[122] When Harman invokes the *uselessness* of the orchid while simultaneously highlighting its beauty, he is drawing out a critique of anthropocentrism, for the forest-dwelling Lingis describes invokes an *originary* aesthetic experience, which transcends human instrumentality and has value and a thing of beauty. Here, Harman can give no reasoned proof to validate this reading, and it is not the case, as in *naturalistic ethical realism*, that we can understand the "good" and the "real" as inseparable from *facts* drawn from the natural world. Rather, as we demonstrate in Chapter 4, this type of attitude or *attunement*, concerned with *Nature* and ethics, is better understood in terms of what Harman has identified in relation to our participation in "aesthetics" as *justified untrue belief*.

We must stress that because of the depth, complexity, and richness of so many of Lingis's descriptions, the possibility exists to draw out a multiplicity of readings that compete and even coalesce; indeed, it is possible to tease out another eco-phenomenological *theme* from the cited passage, which is undoubtedly related to Harman's interpretation, namely, the *ontological concern*, or *solicitous care*, that we *should* demonstrate toward the entities, things, and objects that comprise the living world of *Nature*, which is revealed in a moment of *attunement*. From Lingis's description we are struck with the intuitive understanding that the continued flourishing of our natural spaces *should* inspire thinking and subsequent comportment in the "eco-friendly" direction of striving to protect *Nature* from the impending danger of human encroachment as expressed in ecological concern for the practice of mass deforestation. Here, an ethical vision reveals the human's potential responsibility to *Nature* as its caretaker, as opposed to assuming the contrary role of its superior and master. It is possible, as this interpretation reveals, and we later argue, to connect *ontology* with *axiology* in the quest for an eco-phenomenological ethics. From a biological perspective – and this view tempers and in the extreme obscures the transformative force of Lingis's poetic description, rendering its ability to aesthetically or ethically move us ineffectual – the orchid is parasitic for its nutrition and sustenance on soil fungi and the trees that surround it, and lacking chlorophyll essential for the process of photosynthesis, relies on bees as pollinators for its reproduction by facilitating the production of seeds. It in turn provides the bees with nutrition in the form of its nectar for the feeding and thriving of their offspring. However, as is consistent with our reading of Harman, such a biological explanation, although providing us with an instance of "knowledge" (*justified true-belief*), is reductive and incomplete, and so misses the supreme ontological depth, which shelters the realm of the *axiological*, both at an exterior and interior level, that Lingis's aesthetic and poetic description captures and communicates in non-literal language, through a form of *justified untrue-belief*, which *Nature* selectively reveals within Lingis's passage, in varying modes of *unconcealment* as grounded in the more primordial phenomenon of *concealment*.

Notes

1 Sean Esbjorn-Hargens and Michael Zimmerman, *Integral Ecology: Uniting Multiple Perspectives in the Natural World* (Boston: Integral Books, 2009), 25.
2 Ibid., 23.
3 Ibid.
4 Ibid., 25–26.
5 Ibid., 27.
6 Ibid., 480.
7 Neil Evernden and L Leslie, *The Social Creation of Nature* (Baltimore: Johns Hopkins Press, 1992), 162–163.
8 Esbjorn-Hargens and Zimmerman, *Integral Ecology*, 29.
9 Ted Toadvine, "The Primacy of Desire in Its Ecological Consequences," in *Eco-Phenomenology: Back to the Earth Itself*, eds. Charles Brown and Ted Toadvine (Albany: SUNY Press, 2003), 142.

10 Ibid.
11 Don Marietta, "Back to Earth with Reflection and Ecology," in *Eco-Phenomenology: Back to the Earth Itself*, eds. Charles Brown and Ted Toadvine (Albany: SUNY Press, 2003), 125.
12 Ibid., 127.
13 Arne Naess, "The Shallow and the Deep, Long-Range Ecology Movement: A Summary," *Inquiry*, 16, no. 1 (1986): 95.
14 Graham Harman, *Immaterialism: Objects and Social Theory* (Cambridge: Polity, 2016), 120–121.
15 Paul Edwards, "The Critique of the Cosmological Argument," in *Philosophy: Quest for Truth*, ed. Louis Pojman (New York: Oxford University Press, 2012), 75.
16 Ibid.
17 Marietta, "Back to Earth with Reflection and Ecology," 127.
18 Graham Harman, *Speculative Realism: An Introduction* (Medford, MA: Polity Press, 2018), 117.
19 Graham Harman, *The Third Table/Der dritte Tisch* (Ostfildern: Hatje Cantz, 2012), 12.
20 Graham Harman, *Towards Speculative Realism* (New York: Zone Books, 2009), 131.
21 Ibid., 106.
22 Ibid., 107.
23 Ibid.
24 Harman, *Object-Oriented Ontology*, 68.
25 Harman, *Towards Speculative Realism*, 3.
26 Harman, *Object-Oriented Ontology*, 154.
27 Harman, *Towards Speculative Realism*, 125.
28 Ibid., 128.
29 Harman, *Object-Oriented Ontology*, 153.
30 Graham Harman, "The Future of Continental Realism: Heidegger's Fourfold," *Chiasma*, 3, no. 1 (2017): 84.
31 Ibid., 87.
32 Ibid.
33 Ibid., 86.
34 Ibid., 90.
35 Ibid.
36 Martin Heidegger, *Being and Time*, trans. R. Macquarrie and E. Robinson (New York: Harper-Collins, 1962), 269/227.
37 Harman, *Towards Speculative Realism*, 2.
38 Ibid., 92.
39 Harman, *Immaterialism*, 6–7.
40 Ibid., 7.
41 Ibid., 29.
42 Ibid., 8.
43 Ibid., 12.
44 Ibid.
45 Alphonso Lingis, *The Imperative* (Bloomington: Indiana University Press, 1998), 49.
46 Harman, *Immaterialism*, 40.
47 Harman, *Towards Speculative Realism*, 51.
48 Ibid., 13–14.
49 Harman, *Object-Oriented Ontology*, 167.
50 Ibid., 54.
51 Ibid., 161.
52 Ibid., 155.
53 Harman, *Towards Speculative Realism*, 128.
54 Harman, *Object-Oriented Ontology*, 111.

55 Ibid.
56 Ibid., 150.
57 Harman, "Time, Space, Essence, and Eidos: A New Theory of Causation" *The Journal of Natural and Social Philosophy*, 6, no. 1 (2010): 14.
58 Ibid., 5.
59 Harman, *Object-Oriented Ontology*, 112.
60 Ibid., 163.
61 Ibid.
62 Graham Harman, "On Vicarious Causation," *Collapse II*, 2, no. 1 (2007): 92–93, 171–205.
63 Harman, "Time, Space, Essence, and Eidos," 4.
64 Harman, *Object-Oriented Ontology*, 167.
65 Harman, *Towards Speculative Realism*, 114.
66 Ibid.
67 Harman, *Immaterialism*, 42.
68 Ibid.
69 Harman, "Vicarious Causation," 199.
70 Ibid., 213.
71 Ibid.
72 Harman, *Object-Oriented Ontology*, 115.
73 Harman, *Towards Speculative Realism*, 131.
74 Harman, *Object-Oriented Ontology*, 134.
75 Ibid., 114.
76 Ibid., 133.
77 Ibid., 131.
78 Ibid.
79 Ibid., 137.
80 Harman, "Time, Space, Essence, and Eidos," 14.
81 Ibid., 12.
82 Harman, "Vicarious Causation," 221.
83 Harman, *Object-Oriented Ontology*, 167.
84 Johann Wolfgang von Goethe, in *Conversations of Goethe with Johann Peter Eckermann*, trans. John Oxenford, ed. J. K. Moorhead (New York: Da Capo Press, 1998), 75.
85 Harman, *The Third Table*, 12.
86 Harman, *Towards Speculative Realism*, 137.
87 Ibid., 139.
88 Friedrich Holderlin, "Nature," in *Selected Poems and Fragments*, trans. Michael Hamburger (London: Penguin Classics, 1994), 17.
89 Graham Harman, *Speculative Realism: An Introduction* (Medford, MA: Polity Press, 2018), 121.
90 Ibid.
91 Harman, *Object-Oriented Ontology*, 89.
92 Harman, *Speculative Realism*, 121.
93 Ibid., 121.
94 Harman, *Object-Oriented Ontology*, 84.
95 Ibid.
96 Harman, *Speculative Realism*, 122.
97 Harman, *Towards Speculative Realism*, 139.
98 Ibid., 82.
99 Ibid., 86.
100 Harman, *Object-Oriented Ontology*, 181.
101 Ibid., 258.
102 Ibid., 55.

103 Ibid., 56.
104 Don Ihde, *Experimental Phenomenology* (New York: Paragon Books, 1979), 120–121.
105 George Pattison, *Later Heidegger* (London: Routledge, 2000), 110.
106 Max van Manen, *Researching Lived Experience* (Albany: SUNY Press, 1990), 74.
107 Ibid., 74–75.
108 Charles Brown, "The Real and the Good," in *Eco-Phenomenology: Back to the Earth Itself*, eds. Charles Brown and Ted Toadvine (Albany: SUNY Press, 2003), 16.
109 Harman, *Object-Oriented Ontology*, 247.
110 Terry Barrett, *Why Is It Art? Aesthetics and Criticism of Contemporary Art* (New York: Oxford University Press, 2008), 62.
111 Harman, *Object-Oriented Ontology*, 167.
112 Ibid., 168.
113 Ibid., 185.
114 Ibid., 192.
115 Ibid., 67.
116 Ibid.
117 Harman, *Towards Speculative Realism*, 18.
118 Ibid., 18.
119 Ibid., 21.
120 Karl Jaspers, *Way to Wisdom*, trans. Ralph Manheim (New Haven: Yale University Press, 1954), 20.
121 Alphonso Lingis, *Foreign Bodies* (London: Routledge, 1994), 32.
122 Harman, *Towards Speculative Realism*, 19.

4 Duties and imperatives in a reconceived ethics of *Nature*

Intuitionism and phenomenology

Prolegomena to an eco-phenomenological ethics of *Nature*

We agree with Brown and Toadvine, who recognize the extreme difficulty associated with formulating an ethics of *Nature*, whether in eco-phenomenology or philosophical environmental ethics, for it is essential to avoid reducing *Nature* to the homogenous, the continuous, the predictable, or the thematizable. What is called for initially, and which might serve as the ground for our ethical exploration, is the "rediscovery of our natural experiences as value-laden," and so eco-phenomenology recognizes, as our reading of Lingis highlighted, "nature's axiological qualities as both inherent and ineliminable."[1] Ultimately, the manner in which we attempt to elucidate these issues will diverge from the eco-phenomenological path carved out by these two aforementioned scholars, but we begin with the crucial issues they address in quest of re-conceptualizing a potential ethics of *Nature*. This will eventually lead to our main discussion, engaging both W. D. Ross's prima facie *ethics* and the view of ethical duties linked to "ethical intuitionism" and the phenomenology of Lingis, who makes the unique case, as stated, that fundamental ethical imperatives are present not only within our relationships with other sentient creatures with whom we share the planet, but also, and uniquely, in the very *things of the world themselves*. We understand that when speculating on ethics, eco-phenomenology must point out the futility of adopting an ethics grounded in, as Brown recognizes, moral theories that "share a conception of moral objectivity," incorporating a "single unifying essence" based on the "allegedly single essential character of moral phenomena."[2] This we find in the deontology of Kant, Utilitarianism, Contractarians, and perhaps most prominently, in the moral absolutism of Divine Command Theory. These assumptions give rise to what Brown terms "Rule-Based Moral Monism," where fixed moral obligations are independent of our emotions and desires – ignoring the changing and unpredictable nature of the situations and circumstances within which we find ourselves – and this ideal of objectivity, which is often and quite erroneously linked with the epistemic trustworthiness of the natural sciences, appears to Brown to be "parasitic on the notion of objectivity developed to support the realistic metaphysical interpretation of *res extensa*."[3] Eco-phenomenology must be critical of views that adopt a hard-line ethical view in which duties and obligations are based on *agent-independent* or *agent-neutral* precepts and principles. Since, as Brown makes clear, "everyday experience resists being forced into monistic models," and indeed when considering phenomenology's pre-theoretical realm of "lived experience," an understanding of *eco-ethics* begins with the "initial acceptance of the idea that the prospect of reducing all moral phenomena to a single criterion [is] hopelessly flawed."[4]

We begin our move into eco-ethics by examining two interrelated issues in the history of philosophy that all forms of ethics undoubtedly have to confront or at least acknowledge when attempting to formulate moral arguments: The move from the *is* to the *good* and then the subsequent all-important move from the *good* to the *ought*, which emerge from G. E. Moore's *naturalistic fallacy* and David Hume's *fact/value distinction*. We also encounter, within these issues, the *normative* concern with values and the *ethical* concern with imperatives, duties, and obligations for behaving in a moral manner. Let us first focus on the latter issue of *normativity* in its relation to *axiology*, or for our purposes, the branch of philosophy that is specifically concerned with "values" and "value-judgments." *Axiology* is derived from the Greek, "*axios*," which indicates a concern for those things that have value or *should* be valued, i.e., those things *worthy of calling us to value*, and based on our concerns in this chapter, it is the goal of *axiology* to seek to determine the origin, makeup, meaning, origins, and characteristics of values along with the criterion/criteria for assessing the arguments we formulate in defense of the values we adopt and endorse. *Axiology* includes the study of "goodness," and is most commonly known for its exploration of ethics but perhaps a bit lesser known for its concern for understanding beauty and "aesthetics." It is important to define "normative," for we are perhaps most familiar with the term when referencing "normative ethics," a philosophical field whose moniker appears redundant or tautological. When things are classified as *normative*, this refers to "that which is (a) regulative of human conduct (according to an ideal, norm, or standard), (b) preferential (expressing felt attitudes, values, and biases), and (c) prescriptive (expressing a command or an obligation)."[5] As every first-year philosophy student knows, arguments related to things that are *normative*, which indicates that they are ineluctably bound up with *value*, cannot be expressive of propositional certainty, which is why it is correct to say that when defending one or another ethical or aesthetic position, we couch our claims and supporting evidence in the form of *normative statements*, and again, not in the form of propositions – i.e., neither through *analytic* nor *synthetic* propositions – and most certainly not in the form of valid and sound deductive syllogisms. This is because *normative knowledge* cannot disclose "truth" in the same manner as mathematics or the natural sciences, and here it is helpful to recall what we introduced earlier regarding the type of "understanding" that was linked with phenomenological description, and let us also recall the type of "knowing" that Harman introduced when analyzing metaphor – or aesthetics – which he interestingly linked ontologically and epistemologically with philosophy, namely, *justified untrue-belief*.

Hume frames the *fallacy of deriving ought from is*, which is grounded in the claim that we cannot draw values from statements of fact, or more directly related to our concern, we cannot legitimately derive a moral *ought* from the *facts* associated with events of our experiences. We cannot formulate statements expressing and validating our ethical obligations by focusing on or observing facts about the world. This fallacy is also revived in the analytic philosophy of A. J. Ayer within his view of *emotivism*, which again philosophizes the intractable relationship between facts and values, in that since values cannot be empirically verifiable they must be understood – and devalued – as expressing only emotions. As is typical for classical empiricists, especially the Logical Positivists, who drew from both Hume's empiricism and what might be labeled a slightly overzealous reading of Wittgenstein's *Tractatus Logico-Philosophicus*, only tautologies (*analytic* propositions) and empirical propositions (*synthetic* propositions) are meaningful, whereas value judgments are "meaningless" because they do not fit into either of the aforementioned epistemic categories. In this view, the potential danger exists that if values are nothing but subjective emotive expressions, they cannot be verified, and hence we devolve into a view that is expressive in terms of either ethical subjectivism or relativism or, in the extreme, nihilism, the radical view that no ethical statements can be verified for there is no meaning, knowledge, or linguistic (semantic) veracity in the world. It is G. E. Moore who philosophizes the *naturalistic fallacy*, which we commit when we attempt to discover a notion of the "good" from observing and experiencing natural properties, and this view embraces Hume's original *fact/value distinction*, arguing that any notion of the "good" cannot be extracted from an examination of facts. Here, according to Moore, when we seek justification for our ethical principles, and believe that such justification is possible in the first instance, we wrongly equate ethical statements with factual statements or propositions; and for Moore, this includes the error of deriving "good" from our expressions of emotional "pleasure," as in Hedonism, or "happiness," as in Utilitarianism. In Nolt's analysis of environmental ethics, he brings these positions together to show how environmental ethics must overcome what we have introduced briefly above, namely, the duplicitous problem of identifying the "is" with the "good" and then making the controversial move to the "ought" in relation to the "good." First, there is the "transition from *is* to *good* – that is, a transition from fact to value,"[6] and second, there is a transition from *good* to *ought*, which then allows us, with the understanding that "both kinds of premises are typically controversial," to potentially prescribe "the behavior of agents toward whatever satisfies the factual criterion."[7]

Nolt observes that one way to attempt to bridge the *good-ought* gap in eco-ethics might be through "some clever ordinary language philosophizing," e.g., by demonstrating uses of the term "good" that are bound up with and carry "certain 'oughts' and so establish conceptual, if not logical, links from *good* to *ought*."[8] However, Nolt correctly recognizes that this move ignores the experiential and merely rehearses and reiterates current linguistic habits, telling us little about the "lived experience" of the *normative*. Many philosophical attempts at establishing environmental ethics include appeals to direct obligations to natural entities or obligations to natural entities that follow from obligations to humans or other sentient beings. Indeed, we can also understand the move from the *is* to the *good* to the *ought* through the understanding that values, meanings, and instances of the *good* are always already present and irreducible to our experiences of the world because they are antecedent to and transcendent of our presence in the natural world; we have made this case in relation to Harman's ontology of objects, and as we learn in this chapter, Lingis holds the identical view that values are already present in the *originary* imperatives giving order and structure to the *things themselves*. We argue that eco-phenomenology is unique in that it searches for an ethics by attending to the interrelation between human and *Nature* as it lives as a metaphysical/ontological and axiological phenomenon. As Brown and Toadvine contend, the search for an axiological dimension to our "lived experience" of the world, others, and *Nature* begins with "the articulation of goodness and value within non-human nature, leading us to an experiential, if not ontological, grounding of an ecological ethics."[9] It is within the pre-reflective experience that we experience the world of our involvement and the things within it already infused with value, and thus, according to Brown, our world is "filled with moral sentiments," which instantiate a "prereflective axiological consciousness."[10] This is an awareness or "apprehension," according to W. D. Ross, which we later equate with what we term the *axiological intuition* highlighting our pre-reflective immersed intentional states, within which what is valuable and its valuation manifest as inseparable, i.e., we do not bestow value on the world or *Nature*, rather we encounter it, we reveal what is always already the case. As Thomson observes, this state of pre-reflective immersion in the world transcends both the dichotomies of *mind/world* and *fact/value*, because in this state, which for eco-phenomenology is brought about through some form of "bracketing," according to Thomson, we are able to "think beneath and beyond these conceptual dichotomies, entrenched in our habits of thought."[11]

To introduce the reader to one instance of the eco-philosophical challenge to the *fact/value* distinction as related to those embracing *naturalistic ethical realism*, we explore Brown's philosophy, which offers an interesting version of environmental ethics as a form of *ethical perfectionism*. However, it must be noted that we do not endorse this view *in toto*, because we argue for a revised prima facie *duty* ethics and not a form of *naturalistic ethical realism*. According to Brown, it is possible to escape the charge of "subjectivism" or "ethical relativism" by looking to what he refers to as the "axiological transcendental," which is the "meaningful order of purpose and value" that is inseparable from our "unconscious background of experience available for phenomenological reflection."[12] Through phenomenological reflection we discover that human goods are set within a larger world of goods that is always already present in *Nature*, "arising from the biotic and prehuman constitution of nature as good."[13] On Brown's reading, the moral "good" is linked with empirical conditions that enhance life and contribute to its flourishing, and the moral "bad" is associated with conditions that diminish or impede its flourishing and ultimately degrade and destroy life. Thus, what *is* "good" *ought* to be treated in a certain manner and what *is* "bad" *ought* to call for a specific type of mindset and behavior. It is, as Brown claims, the "meaningfully ordered and value-laden world of our direct experience that ultimately justifies all moral claims."[14] This view that values are intrinsic to our experience of the world is a notion common to phenomenological studies, and here recall our talk of phenomenology in Chapter 1. Detmer claims that we normally view things that exist outside of consciousness in a way that tacitly endorses the *fact/value distinction*, which gives the impression that values are entirely subjective, that they are *projected* onto the world. However, the process of bracketing, we have seen, facilitates "the study of experienced objects as phenomena," i.e., "objects of experience."[15] This undermines overly simplistic and even naïve views to either objectivism or subjectivism, because such "theories are based, not on the evidence of experience, but rather on ideas about what does and does not exist independently of human subjectivity."[16] However, because phenomenology approaches objects of experience as phenomena – as a holistic experience – we find that in our "lived" experiences, values are already present as an irreducible component of our encounters with others, with artworks, with natural landscapes, e.g., we do not encounter a mountain range or stand in the presence of a wild elk in a blank emotionless state; we are always, in one way or another, attuned to its presence through our involvement with it, even if that involvement occurs at a distance, even if our mood is that of boredom.

Thus, borrowing from phenomenology, this is eco-phenomenology's move beyond ethical subjectivism by granting that meaning and value are already present to our most basic dealings with the world. However, the values we reveal and discover are not dependent on human intentionality, and so this view bodes against the idea, as Thomson points out, that "entities only show up within the horizons of [our] concerns,"[17] in terms reducible to an anthropocentric and *instrumental* view of *Nature*, and so in this way, eco-phenomenology offers a reconceived notion of the universal status of ethics. This is because, according to Thomson, when dealing with the pre-theoretical realm of experience, or *immersed intentionality*, it is possible to "distinguish between the entrenched sediments of pernicious traditions, on the one hand, and truly universal values, on the other."[18] This, however, is not to philosophize yet another form of ethical absolutism or "monism" as we have termed it, for as Brown stresses, we must resist reducing morality to prescribing and proscribing rigid and indelible transcendental principles for behavior that determine what actions *must* or *must not* be done at all times despite the situation or circumstance, in terms reminiscent of a *deontological* view of ethics. As opposed to a set of strict and indelible *nomological* ethical precepts, it is possible to draw from our moral experience of *Nature*, a "collection of moral insights and moral tools,"[19] and there is always the possibility present that our moral insights will change and evolve, for it is possible for us to continually assess and reassess the import and weight of these insights. It is the case that the multitude of unique experiences we have and share with others defy reduction to monistic models, and so Brown interestingly suggests that the form of eco-ethics emerging from a phenomenological perspective should be conceived and practiced as a form of "moral pluralism" – and here we are in agreement with Brown – where we might understand and judge that

> a duty to tell the truth may be, in one case, grounded in utility and, in another case, grounded in respect for the person with whom I am speaking. Or, from another perspective, we may find that it is appropriate to have an attitude of moral regard and respect for some nonhuman others because they may be able to suffer and for others because they are components of the biosphere.[20]

The prima facie rule-intuitionism of W. D. Ross's ethics

It is true that Ross melds the ethical philosophies of both Kant and Moral Law ethics and the consequentialism or utilitarian ethics of John Stuart Mill. Ross finds value in utilitarianism's concern with beneficence, but questions the elevation of a single monolithic principle of a hedonistic nature ("The Principle of Happiness") to represent our sole moral duty, for "it can be shown that productivity of the maximal good is not what makes all right actions right."[21] In Kant, Ross finds the idea of justice as a binding ethical principle, but thoroughly rejects the idea that moral duties can ever be absolute and indelibly binding, as in Kant's demand that we adhere unflinchingly to the supreme moral principle of the Categorical Imperative, as expressed in four formulations: the *kingdom of ends*, principles of the *law of nature, ends, and autonomy*. However, it is also accurate to point out that prima facie *ethics* combines what Ross – a famous and reputable Aristotelian scholar – deems the most crucial elements of the systematic ethics of Kant and Aristotle (virtue ethics), which as we develop in detail, are traceable to the notion that certain "potential" ethical duties clearly exist, and the way these duties are recognized, embraced, and enacted are contingent upon the unique and changing circumstances within which the moral agent finds herself. For as we understand from Aristotle, *praxis*, as distinguished from *physis* (nature), is the unpredictable and practical realm of ethical activity where *things can always be otherwise*. Below, we focus on the following three fundamental aspects of Ross's ethics as related to our developing understanding of eco-ethics and phenomenology, which will include a discussion of Ross's understanding of the "moral good" as it is separated off from the "rightness" of actions:

1 We examine the notion of *ethical intuition* or what Ross terms the unique manner in which we "apprehend" prima facie ethical correctness for behavior, which he claims is *self-evident* and cannot be validated in the same manner as propositional knowledge. After we intuit the correctness of ethical actions, we then move to reflect upon them in order to propose a potential and tentative set of prima facie duties as related to our interaction with *Nature*;

2 We also explore the notion that the seven prima facie duties that Ross espouses establish a form of *ethical pluralism* that stands beyond monistic and absolutist moral schemas embracing a single, overarching and essential principle; and, finally
3 We analyze the prima facie duties in relation to (2) to show that they are not absolute and that they can be overridden by other competing duties depending on the unique situation within which the moral agent finds herself, thus demonstrating the potential to overcome moral conflicts when and where two duties clash.

All of this necessary preparatory work will pave the way for our all-important move into eco-ethics and the novel notion, which we cull from the phenomenology of Lingis, of ethical imperatives that are transcendent of humans and are inherent in and imparted to us through our *immersed-intentional* interaction with the *things of the world*, i.e., through *originary* acts of perception.

To use the term prima facie in relation to both ethical acts and potential duties gives the initial impression that we are "speaking only of an appearance which a moral situation presents at first sight and which may turn out to be illusory,"[22] whereas Ross contends that we are able to "apprehend" or *intuit* insights into ethical acts in relation to the self-evidence of such actions, and these initial insights are trustworthy and eventually lead to proposing and organizing a tentative categorization of prima facie duties. For Ross, this indicates that we "apprehend" the rightness of certain acts and then reflect on this "intuitive understanding" in order to establish a prima facie set of ethical duties with respective imperatives that guide and direct behavior, which attunes "our mental attitude towards particular acts,"[23] and this is related to Ross's notion of "moral goodness" and the character and disposition of the moral agent. Such acts as keeping promises, demonstrating gratitude for acts of kindness or compassion, and upholding an attitude of justice and equity are ethical in nature, and in the presence of such actions – whether as author, recipient, or observer – the self-evidence of their ethical veracity is in no "need of proof, or of evidence beyond itself."[24] Although our claims to the "rightness" of acts cannot be defended in terms consistent with empirical or analytic "truths," Ross insists, as might be related to what Harman terms *justified untrue belief*, the ethical rightness of the act of promise-keeping is *self-evident* and so holds the same type of veracity as found in our inferences, although we must be clear that they cannot be defended through the process of discursive reasoning. Importantly, as might be related to Harman's linking of aesthetics with philosophy, Ross equates the type of "knowing" – or we could say "non-knowing" – associated with ethics with the insight we glean from our engagement with aesthetics, both *normative* concerns as we have introduced, both present their insight in terms that are *self-evident*. Thus, according to Ross, the manner in which we "apprehend," *intuit*, or glean and judge the rightness of an ethical act is the same way we *intuit* and so judge "the beauty of a particular natural object or work of art."[25] In both of these instances, their so-called "truth-value" cannot be established or validated by working through premises, the truth of which leads us to the apprehension of a conclusion that can be logically validated.

Although what follows might appear to represent an instance of hairsplitting, it is necessary to address the issue of "self-evidence" in relation to prima facie duties: Interpreters of Ross often argue that Ross claims that the seven prima facie duties are *self-evident*. On the surface this claim as related to a general reading of rule-intuitionism appears unproblematic; however, on closer investigation it fails to provide adequate detail as to exactly what Ross's position is on this matter. For example, Ross asserts in terms that indicate quite clearly that the duties themselves – as overarching and "potential" (because they are never absolute) principles for behavior – actually emerge in a way that resembles Aristotle's understanding of *epagoge*, or inductive arguments as *epaktikoi logoi*, and this should not be too surprising, considering, as we have stated, Ross's role as an Aristotelian scholar. As Aristotle writes, induction is "the leading on from particulars to the universal [*katholou*] and from the known to the unknown."[26] This is closely approximated in Ross's description of how we move from specific instances of "similar" intuitions drawn from many unique circumstances to a general grasping or gleaning of an overarching *principle*, and we know that Aristotle claims that this occurs as we move from individual/particular "perceptions" to an "experience" of those collective perceptions that share marked similarities. Let us listen closely to Ross regarding our gleaning and arriving at the prima facie duties from the antecedent self-evidence of the prima facie "rightness" of the acts we encounter and their subsequent impression on us, the manner in which they weigh on us ethically:

> [W]e see the prima facie rightness of an act which would be the fulfillment of a particular promise, and of another which would be the fulfillment of another promise, and when we … think in general terms, we apprehend [intuit] prima facie rightness to belong to the nature of any fulfillment of promise. What comes first in time is the apprehension of the self-evident prima facie rightness of an individual act of a particular type. From this we come by reflection to apprehend the self-evident general principle of prima facie duty.[27]

Ross is clear that it is through reflecting on the many original *perceptions* – or for our purposes, *intuitions* – we have in the differing circumstances that give way to an *experience* consisting of similar *intuitions*, and then reflecting on the perceptions that comprise that experience, that we glean what Ross terms the overarching prima facie duty, e.g., the duty to promise-keeping, fidelity, justice, and the like. What we are able to state, and here we want to remain as true to Ross as possible, is that our original apprehension or intuition of the appearance of fidelity in the context of our immersed involvement with the world and others *already* harbors and shelters, in an immanent and then emergent form – in its *originary* form – the prima facie duty – for what Ross argues is that in the moral situation, the context of involvement, an essential element of the duty's nature presences or emerges, "though not, as duty proper does, arising from the whole of its nature."[28] What is crucial for our reading is that, prior to any type of cognitive reflection – and here the phrasing glides into a mode of discourse common to phenomenology – *we are already in the immanent presence of the potential duty and the imperative that it should give structure to and guide our behavior*. Since it is the case that we can never give categorical, or grant objectivist, certainty to any of our normative statements, there is no way, beyond the antecedent encounter with self-evident "*prima facie* rightness," to provide the type of verification establishing the unquestioned veracity of the prima facie duties, or should it be more properly stated, give definitive justification for the power they have to attune us, inspire us, and move us in the direction of ethical behavior.

Here, we confront the radical skeptical critique related to not only the impossibility of morality, which is *nihilistic* in nature, but also to the critique of the notion of "intuitionism" in ethics, which, as the case goes, does not display the type of veracity required to defend our ethical claims from a persistent skepticism that wrongly seeks an ethical veracity of our intuitive criterion for prima facie *ethics* that rises to the level of proof that eludes any *normative* statement we might formulate. We must reiterate that Ross admits that neither prima facie ethical acts nor duties are translatable and so remain ineffable in terms of propositional discourse. Recall again Harman's notion of *justified untrue-belief* and the epistemic critique and prejudice associated with both aesthetics and philosophy, namely, the erroneous and reductive view that if we cannot have knowledge that is literal, formable, it is therefore insubstantial, and beyond, useless, a view, according to Harman, based on the false assumption that "*everything that exists must be able to be stated accurately in literal propositional language*."[29] As Harman stresses, this gives the false impression that "nothing can be real unless we are able to refer to it in an accurate prose statement that conveys literal properties of the thing in question."[30] Built into this critique is the epistemological prejudice that not only is there a sharp distinction between forms of cognition, but there also exists a hierarchical structure determining and privileging the superiority of the "modes or faculties" of the Cognitive/Rational over the Affective/Creative, where this latter realm of world-disclosure is linked with *intuition* as we understand it. Philosopher David Best, with his notion of *intuition* as a legitimate form of *emotional cognition*, as we will see below, makes a case against views of *intuition* that relegate it to an unthinking and inconsequential mode of world-disclosure, where *intuition* is devalued because it is perceived as nothing beyond "a matter of subjective feeling, in the sense of 'direct' feeling, 'untainted' by cognition, understanding, rationality."[31] This specific type of classification feeds into the false division embracing the "necessary opposition between on the one hand, feeling, creativity and individuality, and on the other, cognition and reason."[32] Ross never makes a detailed case in defense of his notion of "apprehension" that he introduces into his ethics, and so we attempt to elucidate and defend the legitimacy and trustworthiness of *intuition* in *normative* matters where we are concerned with a prima facie form of eco-ethics.

Separating prima facie *ethics* off from both Kantian ethics and consequentialism, Ross stresses the plurality of his ethics, where the seven duties he identifies defy unification under a single, monolithic, and objectivist moral principle. Ross, however, avoids the charge of subjectivism or relativism because of the self-evident origin of the duties, stating emphatically, "There is nothing arbitrary about these prima facie duties. Each rests on a definite circumstance which cannot seriously be held without moral significance."[33] Since as we have seen, Ross is clear that a sense of finality or completeness eludes the duties or principles as he envisions them, for he insists that the list of duties is "correct as far as it goes though not necessarily complete,"[34] the potential exists to amend Ross's initial formulation of these duties, which are as follows – with the understanding that we draw directly from Ross's philosophy and note that these seven duties are variously identified in the work of others analyzing Ross:

1 Duties to *fidelity* (duties to avoid lying and to uphold our promises and commitments).
2 Duties to *reparation* (accepting the responsibility for addressing harms we have committed).
3 Duties to *gratitude* (recognizing and acknowledging the favors and good deeds others have done for us).
4 Duties to *justice* (punishing and rewarding where appropriate in an equitable manner).
5 Duties to *beneficence* (attending to the manner in which we can make others lives better through virtue, intelligence, or pleasure).
6 Duties to *self-interest* (attending to making ourselves better by engaging in virtuous activity or pleasurable endeavors that are intellectually and spiritually uplifting).
7 Duties toward *non-maleficence*, or "not injuring others," which as Ross informs us, is "apprehended as a duty distinct from that of beneficence, and as a duty of a more stringent character."[35]

We must be careful here, for often the error is made of elevating this seventh prima facie duty to the level of a supreme overarching duty, in a manner that vaguely resembles a monistic ethical principle, such as the "Greatest Happiness Principle" of utilitarianism, which would then serve to undermine Ross's claims to both pluralism and the tentativeness bound up with the seven duties. It is perhaps best to think of the duty to *non-maleficence* as a principle which all other duties partake of or participate in, for each in their own way and in some manner or degree could be said to include a concern for avoiding harm to humans wherever and whenever possible. To reiterate, these seven prima facie duties always immerge in specific contexts within specific circumstances when we initially experience and *intuit* the "rightness" (prima facie) of specific ethical acts.

While there is the issue of the rightness of acts emerging within the specific concerns we face within unique circumstances, there is also the issue Ross addresses regarding the conflict that often arises between competing potential prima facie duties, circumstances within which we cannot enact both duties that call us, each in their own manner, with an ethical exigency that demands we honor and uphold our commitment to doing our ethical duty. In moments of ethical conflict – which never arises in absolutist systems such as Divine Command Theory – for Ross, one duty holds the potential to override another and this is expressive of the difference between our *potential* prima facie duties (PPFD) and *actual* prima facie duties (APFD), but here, we must keep in mind, *not* actual in the sense of a deontological duty to a supreme and unquestioned moral imperative or principle. In relation to what we have stated, Ross notes that "*prima facie* duty leaves us with no principle upon which to discern what our actual duty is in particular circumstances."[36] And yet, there are strong reasons for embracing and choosing in the moment of action one duty above another, which is ultimately determined by the ethical context of engagement and the circumstances that give form and shape to that context. When open to the unique circumstances within a specific context of involvement, we allow the circumstances to reveal and dictate the superior moral weight of one of the competing duties, which attunes us to the ethical appropriateness of choosing that duty over the other based on the moral exigency we experience, to which we subjugate ourselves in the moment, e.g., it might be ethically appropriate to break a promise (*duty to fidelity*) in the greater service of human beneficence (*duty to beneficence or general non-maleficence*). Thus, depending on the circumstances, it might also be the case that doing harm in the service of *non-maleficence* is not only appropriate but is demanded in the service of the ethical. The following hypothetical example, while seemingly trivial, will elucidate a scenario that demonstrates the *overriding* character or nature of prima facie duties. Consider the following situation in which each duty possesses ethical weight ("correctness"), but in the moment of action, one is chosen – weighed and judged – above the other, and by means of this "choice" the *potential* status of that ethical duty is transformed, for it has now graduated to the status of the *actual*:

PPFD I: I make a promise to a student to show up at 3:00 pm to assist with her course work (I have a prima facie duty to honor promises). However,

PPFD II: On the way to campus I encounter a lost and injured child (in light of human *beneficence*, I should always help those in distress in order to do "good").

APFD: In this instance, contingent upon *these circumstances*, I choose and establish the *absolute* duty to help the child, which supersedes or overrides the *potential* duty to keep my promise to the student.

In such instances, when seeking to overcome an ethical conflict and the overriding nature of the duties manifests, we are often plagued by an undeniable and legitimate sense of *moral regret*, for even in moments when we think it justifiable to break a promise, "we do not for a moment cease to recognize a prima facie duty to keep our promise, and this leads us to feel, not indeed shame or repentance, but certainly compunction for behaving as we do," and beyond this we also come to realize that it "is our duty to make up somehow to the promisee for the breaking of the promise."[37] Thus, when overcoming ethical conflicts, there is the accompanying feeling of loss for having sacrificed something valuable in the situation in that we chose – albeit out of a sense of ethical exigency – to abandon one of our duties to which we are ethically beholden, and this feeling of regret might also serve as testimony to the potential validity of the prima facie act and subsequent duty. *If*, for example, our dedication to beneficence holds any legitimate ethical weight, and to act in a contrary manner would bring about a sense of regret for not having done so, *then* this feeling of regret contributes to both substantiating the ethical import or weight of the duty in the present moment as well as contributing to our choosing and hence upholding our duty to *beneficence* in future encounters with others. This, in powerful and legitimate manner, substantiates our ethical understanding that the continued commitment to *beneficence* is ethically justified, and in addition, the so-called "veracity" of the duty is reinforced. Moving forward, we will keep in mind Marietta's concern for an ethics of *Nature*, which in line with our claims should neither simply accept a "subjectivist or emotivistic understanding of moral judgment" nor a completely "untestable intuitionist approach."[38] Our goal, in the remainder of this chapter, is to provide compelling argumentation buttressing claims for a reconceived *intuitionist* view of eco-ethics, and to demonstrate that we "can examine and evaluate our volitions and sense of moral obligation without reliance on [a purely] logical argument."[39]

Toward a prima facie phenomenological ethics of *Nature*

Intuiting the normative: The inherent ethical value of Nature

In our unique analysis of ethics in relation to *Nature*, we reveal another form of attunement linked uniquely with a secular view and practice of *ethical intuitionism*, which initially emerges from the theological tradition in *Natural Law Ethics* as in St. Aquinas: "Medieval natural law philosopher Thomas Aquinas argued that we have a special mental process called *synderesis* that gives us general knowledge of moral goodness."[40] Just as Harman convincingly incorporates a secular reading of *occasionalism* in relation to *vicarious causation*, we philosophize a secular view of *ethical intuitionism* in this and the final chapter through a close reading of both Ross's prima facie *ethics* and Lingis's phenomenology. Our encounters with *Nature*, as related to *Rule Intuitionism*, hold the potential to move us – through an *axiological* attunement grounded in *intuition* – by *intimating* what is right (*good*-moral) and wrong (*bad*-immoral), and what we are given within the context of our involvements are ethical imperatives that are available and accessible to any sentient, clear- thinking individuals upon reflection. We argue for the *self-evidence* of such ethical intuitions, which allows us to arrest the seemingly *infinite regress* of ethical skepticism, which includes the critique of both *naturalistic ethical realism* that is traceable to both Husserl and Nietzsche and *transcendental ethical realism* as emerging from "Later" Heidegger. As stated, *ethical intuitionism* initially emerges from St. Aquinas's idea of the "intuitive" mental function and process of *synderesis* that reveals moral knowledge or the natural moral law. This notion of intuition is developed by philosopher Samuel Pufendorf in his version of natural law ethics wherein he claims that we acquire moral knowledge from nature, but not in the sense that nature implants these "plain and distinct notions" directly into "the minds of newborns" – in the analogous sense of Descartes' claim that God provides us with innate, *foundational* knowledge – but rather in the sense that "nature is said to teach us, partly because the knowledge of this [moral] law may be attained by the help of reason";[41] and it is this claim linking reason with morality that greatly inspired Kant. However, Pufendorf importantly adds that it is also the case – and this we might relate to Ross's understanding of "apprehension," but which is *not*, however, to indicate that Ross is a moral law ethicist – that "because the general and most useful points of it [the moral law] are so plain and clear … at first sight [prima facie], they force assent."[42] However, we note that Ross never develops the inner workings or functioning of what we have introduced as *ethical intuition* or "apprehension" of the ethical *rightness* of prima facie actions. No doubt problems exist when claiming that when moved ethically in various circumstances we can, with a certain inexplicable clarity and distinctness, choose correctly because we glean with an intensity and vivacity the *rightness* or *wrongness* of various acts. Indeed, this is precisely what Hilary Putnam suggests – adopting the language of Wittgenstein – when he recognizes that, within some ethical situations, "I find that I have to say: 'I have reached bedrock and this is where my spade is turned.'"[43] Rolston, although not a phenomenologist, addressing the challenges to environmental ethics, observes that the intrinsic value and *normativity* in *Nature* is experienced or intuited and, "like a tickle or remorse, must be felt to be there. Its *esse* is *percipi*."[44] What does this mean and how might this explain why we can, with confidence, choose to perform our prima facie ethical duties?

Let us explore a contemporary view of intuition to learn *why* it is possible to conceive of it as a legitimate form of cognition and *how* it unfolds in order to challenge the skeptical view that such alternative modes of disclosing the world – as we have associated with the realm of the *normative* – fail to demonstrate a sense of "validity" and are hence untrustworthy. As already stated, it is a highly disingenuous move to sharply separate realms of thought into the *emotional* and *rational* and then, further, to apply a value schema grounded in the flawed and controversial metaphysical dualism embracing the hierarchical structure composed of competing polar oppositions, e.g., consider the psychological distinction between ways of "knowing" that are classified as emotional (or *affective*), and deemed inferior, and those understood as rational (or *cognitive*), which have traditionally been embraced as superior modes of "knowing." We are perhaps most familiar with such polar divisions in the history of Western metaphysics – pre-Socratic concerns to be sure – as expressed in the rigid dichotomies between the *superior-and-inferior* designations of Being and becoming, Reality and appearance, Permanence and change, One and the many, and the like. Thus, to many, any talk of intuiting ethics prima facie smacks of a certain "mysticism"; however, as Rudolf Arnheim asserts, intuition is not a "freakish specialty of clairvoyants and artists but one of the two fundamental and indispensable branches of cognition";[45] and importantly, in line with Harman's understanding of *justified untrue-belief*, Arnheim claims that intuition is indeed a unique state of "cognitive" awareness and insight. The extent to which it is possible to draw an analogy between intuition as we are conceiving it and the findings of cognitive psychology is beyond the scope of this present study. However, it is interesting to note the manner in which intuition functions within the processes of human cognition; here, the *affective* is not separated off from the *cognitive*: "Intuition is best defined as one particular property of perception, namely, its ability to apprehend directly the effect of interaction taking place in a field or gestalt situation."[46] What is given to the senses is "forged into a unified perceptual image by the mental power we call intuition"; indeed, all potential knowledge of the "environment and all orientation within the environment begin with the intuitive exploration of the perceptual given."[47] Intuition supplies us with the organization of things in a perceptual field needed to "cognize," e.g., as in vision, intuition organizes the multiplicity of the things we encounter, "objects of various shape, size, and color," and "by means of field processes," provides us with an organized and unified image. Through intuition we perceive or glean "the place and function of each component" as it is ultimately "determined by the structure as 'a whole.' "[48] As might be related to our discussion of prima facie ethical acts and their relation to prima facie duties, although this is certainly not a line of thought Ross pursues, intuition might be likened to the *cognitive* ability to, in various situations, at once grasp the rightness or correctness of ethical acts, while moving beyond our initial perceptions to see, as it were, the *whole* or the overarching duty immanent in the original intuition, i.e., the prima facie duty giving an ethical structure to the *whole*, which is intimated in a fleeting perceptual manifestation.

Further contributing to this line of thought, Louis Arnaud Reid claims that in our experience of aesthetics, moments of "direct intuition" occur, and this for Reid as related to our concerns, is inspired by and happens through stimulation to our senses, our feelings, and our imagination, producing intuitions that convert "new fresh, *untranslated* insight into value,"[49] and he goes on to add that the insights we glean in such aesthetic encounters, or moments of *intuitive attunement*, are at once "conative and affective as well as cognitive."[50] Related to our earlier discussion, David Best, also writing on intuition, points out that intuition is often devalued because it is wrongly reducible to and hence discounted as merely an untrustworthy form of "subjective indirect feeling untainted by cognition,"[51] and this erroneous view is traceable to and emerges from the dogmatic acceptance of what he refers to as "the *myth of cognition*." Against this view, Best argues that intuition is actually an expression of legitimate insight into the things in the world we encounter, and that although intuitive insight fails to rise to the level of propositional certainty, it is nonetheless valuable and of a makeup that holds the potential to inspire and contribute to our formulating "new interpretations and evaluations"[52] – and here we recall Harman's thought-provoking observation that *justified untrue-belief* holds the power, as a mode of trustworthy and valuable insight, *to legitimately inspire a great deal of fruitful work*. With this talk of intuition, we find a potential connection to Aristotle's work on emotional, intuitive perception in relation to aesthetics, ethics, and values, i.e., it is possible to link intuition with *nous pathetikos*, which as Sonia Kaminska reminds us, is also referred to as *nous dynamei* and importantly "consists in receiving forms perceived by the senses and thus enabling the cognitive process."[53] *Nous pathetikos*, as separated off from *nous poietikos* – which is related to the "active mind" – is usually described as referencing the "receptive," resolute, and anticipatory nature of human cognition. Intuition is a phenomenon of *aisthesis*, a sensual, emotional perceptive occurrence, and indeed, what Aristotle terms "practical insight," which is necessary for "good" moral deliberation and choice, manifests the intersection of *nous* and *aisthesis*. Here, we do not apprehend ethical rightness, locating the *mean* of virtue between the *extremes* (*excess* and *deficit*) of vice in the situation of action (*kairos*) through a discursive process (*dianoia*) of reasoning to a conclusion, but rather by means of direct perceptual insight – i.e., *intuition*.

It is certainly the case that the foregoing discussion of intuition encapsulates Harman's ideas from the previous chapter, and the reader is encouraged to revisit our analysis of *non-literal* language and *justified untrue-belief* in relation to traditional modes of "knowing." For, importantly, Harman's notion of *justified untrue-belief* is related to the realm of the aesthetics and philosophy, and his classification of this form of "non-knowing," in terms of a "non-literal" mode of legitimate *cognition*, relates directly to the *cognitive* process of intuition within ethics, an issue of a *normative* nature concerned with value and the ethical "good."

In the eco-phenomenology of Bonnett, the view of the human's experience of the *normativity* inherent in *Nature* is directly relatable to our elucidation of prima facie *ethics*, although his eco-phenomenology does not endorse a prima facie view, and is more closely aligned with, although not identical to, the type of *transcendental ethical realism* introduced above through Thomson's analysis. For Bonnett is dedicated to the understanding of a moral education in terms of an understandable and potentially applicable curriculum, so the search for values must in some way facilitate or lead to a concretization of these values for educational purposes, as his goal is to "challenge the current dominant Western metaphysical climate,"[54] in a way where eco-phenomenology "offers an enriched terrain for the development of moral education and environmental concern."[55] In relation to our understanding of intuition, recall the discussion in Chapter 1 of modes of "understanding" commonly linked with eco-philosophy's turn to phenomenology and the concern with overcoming of the *fact/value distinction* for an eco-ethics, because Bonnett offers an alternative to the knowledge of *Nature* occurring through the "prism of science," which minimizes the role of the subject in the process. In this alternative to scientific knowing, we become "acquainted" with *Nature* in a way that is similar to becoming "acquainted with a person – a direct, intimate, tacit knowledge that *affects* and is capable of engaging all the senses," a kind of *knowing* in which "personal, moral and aesthetic dimensions are imbedded," a knowledge wherein fact and value are "not separated out because things are perceived in their *life*, wholeness and inherent mystery."[56]

As related to our talk of anticipatory openness to *Nature* and the *things themselves*, Bonnett asserts that through bodily contact with the world, through feeling, passion, and emotion, we engage it most primordially by means of "a receptive *sensing* that is less susceptible to abstract generalization and objectification,"[57] and when we do so, we experience it in its own right as *intrinsically valuable*, and not in terms of its relation to our needs, desires, or instrumental purposes. Here, we are called to demonstrate a form of "acquaintanceship" that demands a "certain preparation of heart and mind to receive what is offered," which as opposed to an attitude of volitional technical mastery or a view consistent with scientific objectivism, more resembles a "*dialogical openness* that incorporates a sense of well-being of things themselves."[58] Bonnet contends that we experience the inherent value of *Nature* through the encounter with its radical otherness, mystery, and integrity, and this normative awakening to and awareness of *Nature's* intrinsic worth is an occurrence of *discovering* and not *imposing* values. The notion in both eco-philosophy and environmental ethics that *Nature*'s intrinsic value is neither reliant on scientific or empirical "facts" about the environment nor linked to transcendent "values" that defy all traditional notions of ethics is found in many strands of scholarship, e.g., in addition to the phenomenology of Lingis, who we explore in detail below, we encounter it in the eco-ethics of Holmes Rolston III, who mirrors the view of Bonnett when emphasizing that "what counts as value in nature is not just brought to and imposed on the ecosystem: it is discovered there."[59] Interestingly, as related to our themes, Rolston insists that it is through the *description* and *analysis* of *Nature* that we uncover and experience its intrinsic and essential value, which opens the possibility of endorsing its *systematic rightness*, and in this experience of *Nature*'s "rightness," as related to our discussion regarding the challenges facing eco-ethics, the "transition from *is* to *good* and hence to *ought* occurs," in that we are transported, we are attuned, and we leave science and the factual realm "to enter the domain of evaluation, from which an ethic follows."[60]

When confronted with the critique denying *Nature's* inherent value, Bonnett emphasizes its fatal flaw is grounded in the erroneous assumption that "value" requires a valuing subject, and it is the human being that primarily bestows and projects value on the natural world, and this *directionality*, from valuing human to valued *Nature*, reduces value to fulfilling the needs of the one valuing. This wrongly indicates that value is not and cannot be internal to objects, but rather that value is based on *relational properties*, and further, that only sentient and conscious entities possess values and hence are the only ones that can bestow or project value. Thus, as opposed to intrinsic value we encounter an anthropocentric view of *derivative intrinsic worth*, which might be construed as essentially instrumental, and so this view remains trapped in the traditional metaphysical predicament that eco-phenomenology seeks to extricate us from, namely, the drive to reduce *Nature* to its instrumental value, showing the highly limited concern "for what it does for the valuer, rather than what it is in itself."[61] To return to Nolt, we find this dismissal of the *inherent value in things* in forms of environmental ethics grounded in views embracing the ethical obligation only to natural entities, most specifically sentient beings, which follows directly from the anthropocentric obligation to humans. For example, we encounter such a veiled anthropocentric view in what Nolt classifies as the "Gandhian Approach," the "Humanistic Consequentialist Approach," and the "Aristotelian Approach."[62] Bonnett rightly notes that to challenge the view of *Nature*'s intrinsic value reveals an essential flaw in rational or calculative thought, i.e., because rational thought exhibits and embraces "its own instrumental orientation," it cannot avoid imposing "this orientation on our understanding of nature."[63] This wrongly indicates that our experience of *Nature*'s intrinsic value is unjustified simply because it cannot be given form and voice in purely rational terms, through "literal" propositional or formable discourse. Bonnett concludes that all fundamental "intrinsic values rightly are resistant to this requirement and properly demand that rationality be subservient to what they intuit – helping to reveal them more fully by working *within* their ambience."[64] Bonnett locates the notion of the intrinsic value in *Nature* in the ideas of what he terms "truth" (integrity) and "beauty" (aesthetic), and he ultimately links these with "moral regards," or what we have termed *normative* concerns, because it is possible to say that both these modes of *Nature*'s presencing, based on our understanding of the realm of the *normative*, inspire a deep and enduring – because they demand it – sense of *respect*.

This is also the case for Lingis, and we find this within his phenomenological view of the *fundamental ethical imperatives* that are inherent in and emanate from the "faces" of others and the *things of the world* (*Nature*); our response or rejoinder to the demands of these imperatives also elicits a sense of respect. As Wheeler confirms, Lingis stresses the immediate and unmitigated power of the imperatives we encounter that "guides the formulation of [ethical] law because this force is exterior to the mind and precedes form and formulation,"[65] hence, we are not the sole arbiters of value in the world. For Lingis claims that value is already present in a way that is antecedent to even pre-theoretical and pre-reflective levels of human existence. This is because, for Lingis, things have their own life and orientation, which does not depend on the presence of the human on the scene; and through perception in our pre-theoretical dealings with things, value emerges as a binding force, manifesting in the form of the *imperative*, in a way that importantly precedes not only our rational modes of understanding, but also what might be termed our phenomenological sense of *subject-hood* and the accompanying emergence of human autonomy. Lingis's phenomenology gives us, in addition to the view toward the sense of ethical *rightness* and *wrongness* as linked to *Nature*, a unique vista into the issue of "moral goodness," or *ethical attunement*. As Wheeler makes clear, in a way that relates directly to Ross's own view as presented earlier in the chapter, for Lingis, "the situations we encounter with other persons and things in the world intrude with their importance, urgency, and immediacy; and these things themselves also imperatively direct and govern our appropriate responses," for it is the "situation itself that requires action, with a force that intrudes with the imperative urgency of what one has to do."[66]

Although the view of intrinsic value in *Nature* we have drawn out does not produce or lead to the establishment of indelible moral, *nomological* principles for *prescribed* and *proscribed* behavior, for "what is important is not always reducible to reason"[67] – and indeed, for Lingis, the "rational does *not* equate with the required"[68] – this is because it is the "imperative force of situations, and not their rational or principled form, that commands my response."[69] Indeed, in these cases, as Wheeler makes clear, "rational justifications of our actions may be pointless and may, in fact, falsify the imperative involved."[70] However, we can at least organize this general phenomenological perspective for the sake of clarity as Nolt does in this concise parsing:

1 Whatever has a good of its own has inherent worth.
2 We ought to respect whatever has inherent worth.
3 So: It makes good ethical sense that we ought to respect whatever has a good of its own.[71]

Thus, in response to the issue of *justification* for our experience of the *normative* in *Nature*, and we encounter this view of *Nature*'s intrinsic value in both Lingis and Bonnet, it is within the "lived" moments of our immersed and primordial experience of *Nature* – and the issue of *immersed-intentionality* will be explored in the next chapter – that we simply "feel" and "intuit" either the rightness or wrongness concerning the ways we are behaving toward, or better, the ways that we are orienting ourselves toward and responding to the call of the imperatives of *Nature*. For example, according to Bonnett, it is in these moments that the *self-evidence* of the appropriate ethical stance as consistent with the inherent value or *normativity* of *Nature* becomes apparent as it intimates to us the ethical "sense of not simply how things are, but how they *should* be."[72] So, it is the case that we simply intuit (prima facie) the wrongness and accompanying sense of outrage when encountering – either in the lived experience itself or through our engagement with a vivid description – an egregious and monumental ecological disaster such as we have described in Chapter 2 in our analysis of the Exxon Valdez Disaster, which is an instance of a grave and monumental *injustice* done to *Nature*, an act that instantiates a callous and cruelly *malevolent* attitude that demands that we uphold the duty to offer *reparations* for the "despoliation of a sea shore" as the waves "covered by an oil slick" blanket and choke out the indigenous life forms on both sea and land. Even on a smaller scale, we experience the same intuitive ethical response when we recognize the need and exigency, in light of a duty to demonstrate *beneficence* toward *Nature*, to avoid the "unnecessary trampling of the bluebells emerging through the woodland floor, or the wanton destruction of an ancient oak."[73] Here, the visceral ambiance of value in *Nature* transcends the human subject, while at once calling her in and binding her to an ethical responsibility grounded in *Nature*'s own unique demands, and this is because the *demand* "emanates from the things themselves."[74] It is, as Bonnett contends, the sense of *Nature*'s integrity and our responsibility and dedication to fulfilling our *openness* to the things themselves that inspires our "strong normative measures in experience, particularly when it is strongly grounded, i.e., emplaced,"[75] or set within the intimate *space* and *time* of our dwelling within *Nature* – and more will be said about this notion of *emplaced* living, or *immersed-intentionality* in *Nature* in the final chapter. Lingis also powerfully philosophizes an ethical concern that is ineluctably bound up with humans, "things," and *Nature*, astutely observing, "What has to be done, overriding our wants, is determined by the intrinsic importance of what is in danger,"[76] i.e., in light of fragility and finitude, which "makes us able to feel the *mortifying* force of a moral imperative,"[77] we *demonstrate* our deep concern and we are beholden to our duty to care for those things that are in danger of perishing, declining, and disappearing. As Wheeler asserts, it is the understanding of *finitude* that "makes it possible to feel the weight of respect as a restraint"[78] on my own desires and needs: in the face of *finitude* I am released from the grip of my limited subjective concerns. Lingis points out that although it is possible for us to dismantle the ancient temples, in order to "sell the parts for our own enrichment," the more we know and appreciate them, "the more it is evident that to subordinate their existence to our private wealth is to denature them."[79] Lingis describes many instances where we encounter in *Nature* an undeniable sense of moral exigency, which finds us in certain situations where we, if we are attuned to *Nature*'s call and address, intuit and come to understand what *has to be done* and so we *do*, with an unquestioned and unwavering sense of *fidelity* to honoring *Nature*'s imperative. In the following passage, which lives poetically as a *normative* description of the sequoia trees, Lingis reinforces our understanding of the undeniable inherent worth of *Nature*:

> Although someone with enough capital and equipment can cut down the sequoia forests for his own profit, the more we perceive and know these forests the more evident it is that the living strands of sequoias have intrinsic importance as an awesome living spectacle of beauty and grandeur, belittling but also inspiring us, but even more that to subjugate their rare great life and great age to the puny pleasures a human will derive from cutting them down for private profit is a perversion.[80]

Thus, as Lingis rightly observes, when we are strolling through the great sequoia forest and come across "a discarded, still smoldering, cigarette butt in the dry leaves,"[81] there is an ethical intuition, an attuned understanding, of what must immediately be done, and we, without hesitation, snuff it out with a hiking boot in a manner that affirms what *Nature* gives us as an imperative to which we are beholden, to which we demonstrate an unwavering *fidelity*, i.e., "the layout of things of intrinsic importance shows what has to be rescued, protected, repaired, restored, or nourished."[82] In this instance, in response to *Nature*'s call, our actions and their ethical correctness assist in averting a potential *malevolent* and catastrophic ecological disaster. It is also the case that embracing and being attuned to the intrinsic value of *Nature* provides viable ethical substantiation on a practical level for us to dedicate ourselves to the fight to eradicate, in terms of potential duties to *justice*, *beneficence*, and *non-malevolence*, what are commonly referred to as "environmental crimes." For example, in addition to the immorality of poaching protected animal species, it is inherently wrong to trap wild animals for profit, which heinously includes the capture and sale of rare endangered species. It is also inherently unethical to indiscriminately pump fecal matter and toxic waste into rivers and lakes, our precious life-sustaining waterways, which has the net effect of killing the river's inhabitants and local wildlife, destroying the indigenous flora as the toxins from the river seep into the soil contaminating the food chain. In these instances, for Lingis, in light of *Nature*'s vulnerability, fragility, and ultimately, its *mortality*, we find ourselves in the grip – the *attunement* – of what he terms "mortification," which is linked with the generation of the "negative feeling of distress" and, in relation to our discussion of prima facie intuition, the intense "suffering [we] feel within [our] sensuous vitality is the weight of evidence [we] have of the force of the [ethical] law weighing on [us]."[83] To reiterate, Bonnett importantly emphasizes that our intuitive ethical responses arising in specific circumstances emerge from the "character of the emplaced individual" in relation to *Nature*, and despite these responses failing to "produce some sort of moral code," they "elucidate one of the *pre-conditions* of a properly informed ethical response."[84] Further, we must stress that intuitive ethical responses are not merely fanciful or imaginative – recall our earlier discussion of intuition – because we do have for ethical guidelines clear and stark examples of what type of behavior toward *Nature* is both ethically correct and incorrect.

It is through our contact with the other, as Wheeler points out, when recognizing and being moved by the other's *need*, her "pain and vulnerability," that "morality is constituted in mortality," or the extreme instantiation of finitude and the potential of extinction, creating a moment wherein the phenomenon of *mortification* "is immediately felt in respect"; and this presses hard on our "sensibility and restricts our desires," keeping our hedonism in check.[85] It is our response in terms of "care and empathy," as Wheeler contends, that is awakened in the presence of the other's *mortality*, for "their mortality *mortifies me*," and this demands and inspires a moral response. As opposed to a religious or theological understanding of *mortification* linked with radical asceticism, in Lingis's phenomenology it indicates subduing and sublimating our desires in service of the other's desires. This subduing and sublimation also represents the understanding of "need" in relation to *Nature*'s ethical imperatives and our exigent duty to those things that demand because they are in *need* of care, in that the others' *need*, the *needs* of *Nature*, as expressed through the *fidelity* to our duty to imperatives, place binding demands on us. Extending beyond just the human face as in the phenomenology of Levinas, Lingis makes a far more radical ethical claim that to understand what non-human "things" are is already to *see what they need and understand what we can do to respond to this sense of need that we intuit.* As Lingis reasons, when we see and experience a tree, we understand that it needs a rich and fecund space within which to grow and flourish; when we observe a fish, we are aware that it requires clean water within which to live, swim, and thrive; and, in the extreme, when we experience and are attuned to the planet in the thralls of immanent destruction, we are *mortified* out of a deep sense of *respect* and moved to respond to its desperate *need* and call for rescue, assistance, and support. This, as Wheeler observes, is Lingis's astute and acute awareness of and need for "advocating the primacy of moral finitude."[86]

Fundamental ethical imperatives given by the world and the things themselves

In this section, we work to elucidate in greater detail the notion in Lingis of "imperatives" in the *things themselves*, or as is our concern, in *Nature*. The importance of imperatives in Lingis cannot be overstated, for as Wheeler confirms, it is through our interaction with the imperative, that we first can "determine what is important in life itself, what intrudes with imperative urgency, and what calls for our response and constitutes our responsibility as mortal and moral subjects."[87] As is well known, and as pertinent to Ross's prima facie *ethics*, the deontological moral law ethics of Kant focus on the crucial notion of "imperatives," and for Kant there are two types: (1) Hypothetical imperative(s) and (2) Categorical imperative(s). Just as Ross rereads Kant and pushes his thought into new directions, so too does Lingis with his unique phenomenological notion of ethical imperatives, which are antecedent, because they are grounded and emerge in the *originary* phenomenon of perception, to the primary faculty that Kant endorsed as grounding the moral law, namely, the faculty of human reason. It is the case, as Wheeler recognizes, that although Lingis grants reason its just due, as he does not devalue it in the quest to privilege the mode of the *affective*, when it comes to ethics, over and against Kant's emphasis on reason's form as universal law, the *originary* force of the imperative is "not reducible to law because it is unformulatable in law."[88] In contrast to existential phenomenology, the world, for Lingis, is not simply "given" through perception as a matter of *ontological certitude*, but rather it is first opened to us in the form of imperatives to which we have a duty to respond, and it is, as Wheeler asserts, through the organizing and "unifying aspects of imperative passion" that the emotions called forth "supply motives for moral evaluations,"[89] and as Wheeler also recognizes, it is this move to introduce the original organizing power of the imperative that traditional "phenomenology has lacked."[90] We are responding to the world's call or address in the act of perceiving it, e.g., Lingis argues, returning to a theme broached in Chapter 1, that when we "wander through the azalea plants in bloom, our movements are not spasmodic reactions to clouds of atoms, nor to masses of sensations scintillating in our minds,"[91] but more accurately, our walk, its gait, is a measured response to what we encounter in our perceptual field, in response to the things of the world that, as we have already seen, are oriented in their own and *owned* unique ways. Lingis is emphatic that there is a "normative coherence, a consistency in the evolving zone of perception,"[92] a view which is importantly relatable to our current concerns.

The things in the world we encounter and engage with, according to Lingis, carry with them a "body of knowledge" or a "competency, a praktognosis, a knowledge of how to deal" with things.[93] We must pause briefly to examine this term "praktognosis" in Lingis, this because we have examined Heidegger in some detail, and especially considering what we learned about the manner in which Heidegger's phenomenology was reconceived by Harman. Praktognosis is not the equivalent of "practical know-how," as would be consistent with analytic and what are usually erroneously termed "pragmatic" readings of Heidegger. Although it is the case that Aristotle in *Posterior Analytics* links *gnosis* (*γνωσις*) with *episteme*, in terms of "scientific knowledge," "praktognosis" might be employed to reference a sensual form of "recognition" emerging from *aisthesis* and facilitating and informing our intuitive enlightenment.[94] Thus, the term Lingis uses is meant to represent neither an a priori form of "knowledge" nor an empirical "knowledge" linked with the sciences, but rather this "insight" as *gnosis* emerges when we position or situate ourselves before and mold ourselves to the things, when we move with and in concert with the things we perceive so as to "pick up the style and organizing schema" inherent in the things that materialize "on the planes and levels of the sensible,"[95] an engagement which indicates that we belong to the world. This notion of *gnosis* as related to *aisthesis* is a bodily or "embodied" form of insight, where things, as Lingis observes, "induce in us an appropriate posture and itinerary of discovery," e.g., when encountering a "great fir on the crest of the mountain, we stand tall and our eyes travel upwards to the clouds and eagles."[96] Thus, when encountering things – and recall that this experience is *normative* – we take into our postural schema the internal lines, the tensions and orientations, of the things we experience. And when we orient ourselves to the "motor diagrams" of the things we perceive, with the "visible, tangible, and audible form of ourselves turned to them,"[97] we conform our posture to them, attuned by the imperatives emanating from them.

In Lingis's view, then, we see that the normative is inseparable from our perception of and immersion in the world, and we explore the notion of *emplaced intentionality* in the next chapter. According to Lingis, our subject-hood is contingent upon an *originary* ethical encounter with the *normative* and gathering force of the imperatives we encounter in others and things, and when orienting ourselves to the demands of the imperatives we are *subjecting* ourselves to, through an act of *subjugation*, their primordial force. It is the fundamental force of the imperative that weighs on us and reveals the unshakable responsibility we have to its demands, a demand that is antecedent to any rational or formal understanding and communication of the imperative. As Lingis contends, in contrast to Kant and in a way that relates to Ross, imperatives are "formulated by an exercise of thought that is already subject to the imperative," and so the "imperative is obeyed before being formulated"; indeed ethical principles can be formulated *only* "because they are obeyed."[98] Here, recall our analysis of Ross, wherein the original prima facie recognition of the rightness of an act is intuited and then the overarching duty emerges only after reflecting on the potential reasons *why* such acts *should* inspire tentative ethical principles. However, we must stress that for Lingis, the essence of the imperative always remains "ungroundable and unrepresentable," and so the ethical understanding we may arrive at, the principle we may attempt to draw from it, along with "every argument with which [we] might try to demonstrate it already presupposes a thought subjected to the imperative";[99] the imperative always precedes and grounds the formalized law. Where we have employed the term "intuition," Lingis incorporates "affective substance," which indicates that our relationship to and the stance we take in the face of the *originary* ethical weight of the imperative is related to emotionality and the mind's receptivity to its emanating, radiating force. Although speaking of emotionality, we must be clear that this does not indicate that our *wants* and *desires* determine our response to imperatives. This is because obeying the imperative and our subjection to it is always "determined by the intrinsic importance of what is in danger and the intrinsic importance of what needs us in order to exist,"[100] or as we have already stated, it is an emotionality in service to the other's *mortality* – the finitude of the other, the potential fragility, vulnerability, suffering, and *need* of the other. Lingis, within the following scenario, reveals how adherence to the demands of imperatives transcends both our personal wants and desires as well as any rational justification we might attempt to provide for such adherence to imperatives. Imagine a mother caring for her brain-damaged child: She does not "first formulate beliefs" and allow these to serve for reasons or justification for her nurturing behavior – and indeed, "she may not even be able to supply reasons for doing so";[101] in addition, she is unconcerned that both her immediate and future personal wants and desires will out of necessity be sacrificed in the diligent care for the child, which demands her adherence to an ethical imperative in this moment. As Lingis claims, "[I]nsights come as gifts from a contact with things where we give ourselves completely to them and they to us."[102] In this instance, to offer but one interpretation related to a potential *ethical (PPF) duty*, we might imagine that within this circumstance the mother's binding imperative to *self-improvement* is overridden by the more urgent duty to the imperative of *beneficence* or even the *fidelity* to care for a loved one.

For Levinas, the face (*le visage*) represents the other's locus of ethicality, which is an encounter that precedes ontology and defies an understanding through phenomenology. Although we associate the realm of the ethical with the other's *face*, with its ability to "speak" in terms of inspiring dialogue, it is not an "object of experience in the sensible sense of the term, relative and egoist,"[103] i.e., it is not really present, but rather instantiates the other that transcends my existence as other. This is the case, as Wheeler points out, because Levinas traces ethics to radical alterity expressed through "the unknowable ethical infinity of God";[104] for Lingis, conversely, morality can "be found right on the face of others," and so, as opposed to "exempting the absolute infinity of moral values from further assessment," Lingis argues that valuation occurs "in the medium of the face,"[105] which generates its own ethical imperative directives. The *face*, as Wheeler observes, is the "elemental ground on which our stand acts as support and ground for the Other,"[106] and Lingis works to bring the "importance of Levinasian alterity back to earth with his rectification of Levinas's metaphysics of morals, via the "elemental imperative.'"[107] Lingis embraces the notion that through our encounters with others values emerge that are either beneficial to life in an affirming or ascending manner or detrimental to an ascending life, and the face emanates these values in terms of imperatives that demand our authentic response. Unlike many analytic readings of Heidegger's *Being and Time* focused on the division between *authenticity* and *inauthenticity*, claiming that these terms are value-neutral, it is possible to make a legitimate distinction in Lingis between authentic and inauthentic responses to the imperatives in the other's face wherein both of these terms carry substantial normative weight – this he calls the difference between the "idol" and the "fetish." The face for Lingis, as the "locus of expression and of valuation," is in its self-valuation a form of *idolization* in response to the *need* of the other, and when we make a mockery of the *idol*, we caricature it and transform it into its opposite, namely, we turn it into a *fetish*.

Nature, as we have stated referencing Harman's reading of Lingis, presents to us as imperative and thus it too has a *face*; even the mountains and trees, masses of "semi-aware organic material," presents their faces to all who encounter them.[108] In light of Harman's OOO and Lingis's phenomenology, the things and objects populating *Nature* are each in their own right imperatives, "radiating over us like a black sun, holding us in its orbit, demanding our attention, insisting that we reorganize our lives along its shifting axes."[109] When we respond authentically, in an ethical manner, to the commands of the imperatives we encounter in Nature, it radiates as an *idol*, and when this happens, as Harman contends, a "reversal occurs," in which the other as an object subjected to the "crushing network of earthly laws and determinations" transforms before us and "it reverses into the other as an autonomous commander, by virtue of the task [it] confronts us with."[110] And yet, the reverse can also be the case, for the potential exists for us to *fetishize* objects, for much like any *idol*, objects hold the "propensity to be transformed into and thus assume the role of the *fetish*," when, for instance, we flee in the face of our duty to the imperative in things. When we fetishize things they no longer drift "across the world like an independent power, but [are] now used to manipulate and enslave."[111] A fetish, as Lingis stresses, is used to serve our *instrumental* purposes, related, as opposed to our ethical duty, to our egoistic needs and wants, "put in the service of one's fears or one's cupidity. The idol is noble; the fetish is servile."[112] Here, in direct relation to our eco-phenomenological concerns and view of the *metaphysics of presence*, to de-fetishize *Nature* or the things we encounter – in the service of the imperative – is to liberate things from such restrictive epistemic views and *naturalistic* attitudes associated with *materialism*, *naturalism*, *scientism*, which are all modes bound inextricably with an anthropocentric worldview common to *secular humanism*. It is possible to state, as related to our analysis of the CF, that *Nature* as a de-fetishized phenomenon is freed from the "en-framing" grip of technology. As related to the notion of *fetishizing* objects, we must be mindful that the world of objects in its essence, in terms of what Harman has referenced as the *thing-in-itself*, can never be captured and exhausted in any form of "knowledge." This issue is ineluctably bound up with our *normative* relation with things of the world – namely, the way we grasp and the way things recede from our grasp when dealing with the *normative*. For we must, in advance, already be of the mindset informed by the drive to avoid *fetishizing* things, open to the understanding that all of our encounters with things is given structure by finitude and the sheer unknowability of things – the *excessive* nature of their Being – because as Lingis contends, "despite our own stock of concepts and our taxonomies," what is "continually implicated" in our encounter with things is that "we are trying out their reality in indecisive and inconclusive appearances," which offer us only at best their essences in "consistent and coherent evidences that elude formulation,"[113] and hence continually outstrip our best efforts to bring them to stand in "knowledge."

As stated, for Kant the categorical imperative has an indelible universal status and in Levinas there is an eternal or infinite nature attached to the ethical imperative, indicating that neither philosopher considers the importance of circumstances and the uniqueness of the situations within which we find ourselves to be of importance or a point of ethical concern. This is not so with respect to Lingis and, as we have seen, Ross's understanding of prima facie *ethics*. For Lingis argues that while imperatives are binding they do not float free above the existential situation of our immersion in the world, and so according to Wheeler's analysis of Lingis, "these situations are contingencies and not principles," which indicates that because these contingencies possess a "commanding force"[114] that is beyond rational explanation and without categorical validity that transcends any and all contingent circumstances, as in "objectivist" ethics (*agent-neutral principles*), there is an undeniable *pluralism* in Lingis's phenomenological ethics. When Lingis introduces the concept of the "conflict of imperatives," which is an issue highlighting the ad hoc nature of imperatives (inherent *overridingness*) in a way that recognizes and embraces their contingent, potential, circumstantial, and situational (*not* in a pejorative sense) makeup, we note important similarities between this idea in Lingis and aspects of Ross's philosophy. For what has to be done, according to Lingis, "requires attention to the concrete particularities of this situation," and the thinking that generates "*what I have to do* is ad hoc."[115] At times, depending on the contingency of the circumstances, it is demanded that I turn to "the one who faces me with his or her joy" and "turn away from those who have no one but me to look to in their pain."[116]

Here, both situations are structured by imperatives, and each imperative holds its own pressing and legitimate weight, but choices must ultimately be made. Importantly, in relation to the ethical dilemma of the *conflict of duties*, Lingis shows us, in a manner similar to Ross's prima facie *ethics*, that the imperatives possess the ethical trait of *overridingness*. So, for example, if I am called to the duty of expending my funds on the medical treatment of my son afflicted by a rare disease, requiring repeated surgeries and demanding around-the-clock attention, I am neglecting the needs of my other children and wife, all the while understanding, though based neither on a discursive explanation nor rational justification, that in this instance, in this situation, the exigency of my son's illness calls me to a duty to subject myself to *this* imperative at the expense of others, for in this contingent situation, this is the imperative that holds superior ethical weight. Just as I cannot at once tend to my land in the time of drought and, as Lingis observes, "go to Ethiopia to assist engineers trying to dig wells."[117] As we have seen in Ross, when we experience the *conflict of duties*, we are afflicted by a sense of *moral regret* when making the choice to adhere to one imperative or duty above another. Lingis also recognizes "moral regret" in the sense that an "imperative is a practical necessity," and when we choose one over another in light of an ethical dilemma, it ultimately "prevails over what is undoable";[118] and what is *undoable* is what must be sacrificed in the service of ethics, because although it is overridden and passed over in the context of decision and choice, it too holds monumental ethical importance and commands our respect, and so weighs heavy on our conscience as a burden.

Moral goodness and attunement

Often, scholarly analyses of prima facie *ethics* gloss over the crucial distinction between moral "goodness" and the "rightness" of acts, for according to Ross, "moral goodness is quite distinct from and independent of rightness, which belongs to acts *not* in virtue of the motives they proceed from, but in virtue of the nature of what is done."[119] So, for clarity, we make a distinction between the fidelity of promise-keeping as a morally "right" act and the essence of the human's or moral agent's character or disposition – its moral "goodness" or immoral "corruptness" – from which such acts are traceable as their origin. For Ross, what is termed the "*morally* good" is understood as "good either by being a certain sort of character or by being related in one of certain definitive ways to a certain sort of character."[120] The moral character of which Ross speaks, based on what we already know about him, is traceable to Aristotle's (as well as earlier ancient philosophers') understanding of the moral agent's *hexis* or disposition to behave ethically, which must be educated, trained, and molded in an ethical direction if one is to rise to the status of *phronemos* or, to paraphrase Aristotle, "the one (man) who deliberates and chooses well in the moment of action based on the right motives." Ross discounts the phrase "ethical motivation" and prefers "desire" as the originating ethical force shaping one's character. In short, we *demonstrate* – here, we avoid "possess" as this wrongly indicates that the moral character, once forged, becomes indelible – that we have a morally "good" *desire* to bring about or show *interest* in the happiness of others, i.e., "sorrow at the misfortune of others is morally good because we think that it springs from an interest in their happiness," from the *desire* that they *should* be happy. With regards to the "rightness" of an action, it is possible to understand moral "goodness" in the following manner: Telling a lie in the service of human *beneficence* – having overridden the prima facie duty to *fidelity* – demonstrates the *desire* to render an appropriate ethical judgment, and so here, in this specific and contingent circumstance, lying (as a prima facie ethically corrupt act) is the morally "good" thing to do, and one of "good" moral character/disposition, one demonstrating the morally "good" desire to spare or arrest unnecessary psychological/emotional suffering, sees this and chooses thusly.

Importantly, against Kant, Ross refuses, much like Lingis, to discount the crucial contribution of emotionality informing ethics. Kant is adamant that to be "subject to passions and emotions is probably always an illness of mind because both emotion and passion exclude the sovereignty of reason."[121] What Ross calls "conscientious actions" spring from a "certain desire (the desire to do one's duty) and owes its goodness to the specific nature of the desire."[122] Along with conscientious actions Ross identifies two other morally "good" actions, which also "owe their goodness to the nature of the desire they spring from":[123] the desire to bring about something that is "good" and the desire to reduce or alleviate the other's pain in the service of "happiness." We have talked of ethical duty and imperatives, and Ross agrees with Kant on the importance of one's "sense of moral duty," but disagrees with the notion that this is traceable to our rational responsibility, for rather it is, what for Kant is anathema, namely, "desire" that *should* be the best and most viable guiding reason for recognizing and adhering to our ethical duty. For Ross admits that his ethics provides justification for accepting that our "instinctive generous impulses" inspire our insight into and adherence to "the operation of the sense of duty."[124] Thus, he stresses that the emotions – i.e., the "operation of warm personal feeling" – are ineluctably bound up in and inseparable from the supreme "moral value of the sense of duty."[125] Here, implied in Ross's ethics is the following hypothetical logic: *If* one is attuned (demonstrates the desire) to both the other's needs and one's ethical duty to recognize and respond to prima facie imperatives – one is, for example, ethically conscientious – *then* one demonstrates a character that boasts a morally "good" constitution. It is possible to state that by continually attempting to hold oneself in what we shall call an ethically *open* and *anticipatory* stance before our duty to imperatives, we are engaged in the heuristic process of ethically developing – in terms of what we have identified as phenomenological *self-hood*, a process *not* a destination – in and through the choices made when attuned to our ethical duty, our steadfast responsibility to imperatives. This understanding of moral "goodness" in Ross and the notion of attunement as the transformation to and development of a "good" character can be related to our reading of Lingis, in that the possibility exists – for recall that *praxis* is the realm where things can always be *otherwise* – for us to approach humans, things, and *Nature* as either *idols* or *fetishes*, which as we have already stated, indicates that in the former case we valorize and honor the thing, while in the latter case that we are complicit in the thing's degeneration, and contra to our ethical duty, bastardize it by reducing it to a pure material object for *instrumental* manipulation. Let us take another look at what Lingis believes occurs when we engage the world as set before us and opened to us in terms of imperatives and the subsequent ethical attunement or change to our Being-in-the-world that is arguably occurring.

The ordering of the world occurring in perception, as we have shown, first becomes possible by the power of the imperative, and it is not simply that we perceive what is "there" standing before us, because in response to imperatives, as Lingis contends, our ethical vocation is revealed as a "telos, a task,"[126] a phenomenon wherein we are faced with a potential challenge to our *subject-hood*, to our consistent development as human beings, for we are "subjected in each thing to a reality in the interrogative mode."[127] The structure and order of things, which is inwardly represented, becomes the "effective directive of our action," and when authentically and ethically attuned we assume a posture that instantiates our supreme "respect for our environment."[128] We know that for Lingis the things of the world and *Nature* possess their own unique orientations that are not determined by us and that in truth determine our responses to them; that we are called to respond to the imperative in *Nature* already presupposes a *normative* experience of the world. Recall our previous analysis of human perception in Lingis, which is a state of immersion in the world, where we encounter things in "their upright positions," through which our body orients its posture in compliance to and in harmony with the things we encounter "in the most centered, dexterous, and forceful ways."[129] Recall also, in relation to our current concern with moral "goodness" and attunement, Harman's talk of the "tree effect" that enraptures and transforms us when in the presence of the majestic sequoias in the depths of the forest. Encountering the trees' upward thrust, when their *faces* present to us and we are receptive to subjecting ourselves to their imperatives, we are in the midst of an authentic experience of the trees as *idols*, and in response to what their *faces* present to us, in terms of imperatives, we are attuned in a way where we momentarily stand outside, as an ecstatic (*ek-static*) event, former understandings of *Nature* tainted or stained by the residue of the reigning CF. For Lingis reminds us that such instances are never encounters with mere "things" as rote material objects, but rather moments that are beyond rationalization, beyond the "concept-generating faculty of the mind," moments when we immediately feel, sense, and intuit the trees' "force and measure" and participate in the "uprightness of their life" as it is taken in, which in turn generates our own inspired "upright aspiration" within our own bodies, our own "vertebrate organism."[130] As we have indicated, imperatives emanate from all things in the world and to respond in one or another manner determines our *subject-hood*, inspiring and facilitating a change to our attunement and the way we respond to the world and others. As related to our reading of Ross, Lingis's phenomenology demonstrates that in such moments when we bow in the ethically attuned subjugation to things as *idols*, we simultaneously *de-fetishize* these things from the oppressive weight of *naturalism*, and there is an ethical "rightness" or "correctness" to such acts; and as we assimilate some of their essence through a receptivity to the call of things, their address, their audible patterns are captured for our own speaking, and thus we speak with a new and renewed voice, for to reiterate, *their* essence is awakened in us, and this is a form of attunement that might be linked with moral "goodness." On this point, we learn in Chapter 5 that our potential ethical response to the imperatives of the world and *Nature*, for Lingis, is inextricably bound up with our propensity and capacity to listen in anticipation for and respond to the call or address of *Nature* itself, as an indispensable component of our developing *self-hood* or *subject-hood*.

Whereas Ross stresses the crucial ethical understanding that we must avoid malevolent actions against others – indeed he asserts that *non-maleficence* is set off from *beneficence*, and is perhaps a "duty of a more stringent character"[131]– Lingis focuses on the *fragility* of all things as the determining factor for what *must be done*, and that is "the intrinsic importance of what is in danger and the intrinsic importance of what needs us in order to exist."[132] However, it is possible to understand this notion of concern and care for what is in danger, what is in *need*, as already embracing Ross's understanding of both *beneficence* and *non-maleficence*; the radical difference, of course, is that for Lingis, as we have learned, this care for things in *need* extends to all things (and *Nature*) and is not restricted to humans, for "realities outside ourselves also present themselves – through the faces they wear and display as imperatives – as having to be spared, protected, or promoted."[133] The entire layout of *Nature* indicates that it has to be "rescued, protected, restored, or nourished," and such action becomes obvious "when we appraise the forces and resources available," which presents situations where our own needs and desires come into conflict with what must be done as instructed and determined by the imperative. In deciding what is to be done, which "postpones or interferes with what [we] want to do and with the urgency of [our] hunger and thirst," we instantiate moral "goodness" and commit to the enactment of actions that are at once ethically "right" and morally "good." However, the ethical "good" in question most certainly relates to humans: The good bound up with caring for our family and loved ones, our patch of land, our sustenance and shelter, the clean water to drink and the unpolluted air to breathe are all things that have to be secured.

Returning to the phenomenological-ontological questions posed in Chapter 1, we might envision the earth in ways that transcend the grip of *naturalism* and the modern drive to *fetishize* the natural world. Because now, and recall our introduction of phenomenology, we can return to conceptualize the world with the enlightenment we have drawn from Lingis's philosophy focused as it is on coming to understand the pre-reflective world of our engagement with the binding imperatives that *Nature* instantiates and emanates, phenomena that underlie, precede, and inform both our theoretical and our practical activities. With this understanding, it is possible to respond in a renewed and ethically authentic manner to *Nature*'s *need* when we experience vast swaths of the landscape in desperate and depleted conditions, for example, when destroyed by the adverse effects of hydraulic fracking in the quest for fossil fuels, which include the leveling of surface areas and the contamination of waterways and groundwater supplies. When we see and internalize the ominous sight of the chimneys of large manufacturers belching out massive clouds of white, toxic smoke resulting from the burning of fossil fuels, and as we breathe in the noxious fumes, we are in the presence of *need*. To witness gallons upon gallons of polluted wastewater being dumped into rivers and waterways by commercial industries is to experience the *need* of the earth and of non-human sentient entities that depend for their survival and flourishing on clean water and fecund ground. To witness the unethical treatment of animals within the mass industry of factory farming is to already feel and intuit that there is an ethical imperative for our response to their suffering that should far outweigh any concerns with the immediate satiating of hunger pangs. To be *as* ethical always requires the choice to be *otherwise*. So, the monumentally weighty question Lingis confronts us with runs thusly: Shall we continue to *fetishize* our fellow humans, non-human animals, and the multitude of precious things we encounter, in and as *Nature*, or will we work to transform our mindset and behavior in order to embrace them, and hence let them serve as our ethical guides, as *idols*?

Revised prima facie duties as tentative eco-imperatives

Intuitionism and phenomenology

Brown presents us with an excellent example of the value or "good" we encounter in *Nature* and argues that when experiencing what is valuable in its own right and "good" we understand what we can expect of it; we also intuit what circumstances negatively affect it and what conditions hold the potential to degrade it, destroy it, and transform it into its opposite. Brown also envisions the "good" as a value present despite our human presence, or whether or not it is humans who might benefit from it and partake in it. Recall our original concern with the ontological life or *originary* presencing of a river when our natural attitude is temporarily "bracketed" or held in abeyance, and consider the "good" that is ineluctably bound up with water, which as Brown claims, is a "good" that is shared. He invites us to imagine a thirsty herd of elephants trudging over long distances and at long last finding water, and then pausing to drink, play, and bathe in the water. He observes that we can understand the animals' embodied appreciation and satisfaction in their encounter with the water's goodness. We too can experience the goodness of water, for in parched moments when drinking cool, clean, fresh water we become aware of its necessity, beyond the mere experience of the pleasurable, to our healthful wellbeing. In both these instances we can at once be attuned to an awareness of the "good" in connection with our experience of things and also envision the specific conditions that would destroy the good: In the former, we can imagine all too real circumstances when the fragility of the good shows its face, when, for example, the elephants that roam free are poached for their ivory tusks, this despite the international ban on the ivory trade. In the latter instance, we can again imagine all too real circumstances where drinking water is scarce or polluted with levels of toxic chemicals that make it undrinkable. In 2015, people in the state of California experienced a water shortage that rose to the level of *severe drought* status.

Lingis writes about what he classifies as "cultural codings," which include a wide array of passions and emotions that manifest as values on the faces of others, and as we have seen throughout, on the *faces* of the *things themselves*, for as related to our thoughts in Chapter 2, the manner in which we respond to things of the world, to *Nature*, holds undeniable cultural and political-social implications. These values, and for us we have seen that they emerge through the force of the imperative, move us and inspire us to action. "It is on the faces of others that we discover their values," observes Lingis, but it is not through discerning the meaning of the values they emanate that moves and attunes us; rather it is through "the figure they materialize, shaped not by the pressures of the world but by the joyous or rancorous inner force in them, that we see what their sensuality idolizes or fetishizes."[134] We now attempt to move from the initial analysis of prima facie *ethics*, while attending to the rich store of phenomenological speculation on *normativity* that we have drawn from reading Lingis, to suggest what we will understand as "tentative" imperatives – or *duties* – for potentially structuring and guiding our ethical behavior in the face of *Nature*. We attempt this with the awareness that it is impossible to draw an ethics from Lingis that lives beyond a tentative state as it emerges from our analysis, for as we have seen, to attempt anything resembling a rigid and binding code of ethics related to *Nature* would betray Lingis's phenomenological project and push our own project in the direction of the strict, monistic, and *agent-neutral* deontology resembling Kantian ethics, which we have taken great care to avoid. We maintain the distinction, in and through our reading of Ross, between the ethical "correctness" or "rightness" of an action and the manifestation of the moral "good," as it is related to an ethical character or disposition that is attuned to the binding responsibility we are beholden to in the face of the imperatives we encounter in the other. In relation to these foregoing remarks, recall what Ross stated about the original seven prima facie duties – namely, that in addition to their status as "potential" duties, which indicates that they can be overridden depending on circumstances, it is the case that other duties might exist that Ross overlooks, and for this reason he believes that those reading his ethics are open to thoughtfully and legitimately amend, change, and extend his original seven duties in light of other and perhaps more appropriate ones. For although they serve legitimately as a conditional set of duties, as a caveat, Ross wisely recognizes that they are "correct as far as it goes, but not necessarily complete."[135]

Let us then consider the following seven prima facie *ethical imperatives or duties* related to our reconceived relationship with *Nature* drawn from our reading of these two philosophers, with the understanding that the inclination toward carrying out these duties lies in our openness to them, to what they emanate, and the "apprehension" or intuiting of the rightness of the actions undertaken when responding to the *call* or *address* of these imperatives or duties. Recall that, for Lingis, imperatives represent ends that serve and function as directives ordering our perceptual fields, and so the essence or "reality of things is not [simply] given in our perception, but orders it as an imperative."[136]

1 Duties/Imperatives to *fidelity* – This is the faithfulness we demonstrate to our obligations or duties toward others and *Nature*. This faithfulness to the obligations and responsibilities we have for heeding and adhering to *Nature's* imperatives is inseparable from our immersion in the natural world, for as we have seen, it is only through and in response to imperatives that we disclose the world and open our potential for authentic *self-hood* in relation to the *normativity* inherent within the *things themselves*.
2 Duties/Imperatives to *reparations* – This is the duty we have to others and *Nature* to make amends and offer recompense for wrongs that have been done or injuries that have been inflicted. The damage done to *Nature* in the name of anthropocentricism, as we have shown throughout, poses the greatest risk and most acute danger to the natural world. In order to respond and subject ourselves to this duty we must be re-attuned to the world and *Nature* as that which exceeds us and so a sense of respect will potentially, as Lingis indicates, arrest the *hedonistic drive* to subject *Nature* to human machinations in the service of rote *instrumentalism*.
3 Duties/Imperatives to *gratitude* – This is the duty to demonstrate and offer thankfulness for what has been given by others and *Nature* as gifts; this is the deep appreciation we feel and must show for favors granted or benefits received. The sense of gratitude that emerges, as an imperative, first becomes as potential when attuned to *Nature* in a way that transcends the oppressive influence of either the *naturalistic* attitude or *secular humanism*.
4 Duties/Imperatives to *justice (equity)* – This is the duty toward others and *Nature* to demonstrate fairness and adhere to a sense of rightness and equity within all of our dealings. It is possible to recognize that there is already a sense of *justice* in the duty to *fidelity*, which is also related to the conscientiousness we show toward upholding what we intuit as the "moral right."
5 Duties/Imperatives to *beneficence* – This is the duty we have to others and *Nature* of doing "good" by them in upholding the attitude of approaching others and *Nature* with an acute sense of charity, respect, and kindness. This we might say is also demonstrating a sense of deep and solicitous care for the other's continuing and flourishing wellbeing, based on the recognition of the other's *fragility* and potential to experience pain and suffering; we have termed this the "destitution of the other" which elicits in us a shattering sense of *mortification* in light of the overarching ontological condition of finitude and the extreme and terminal possibility of the other's *mortality*.
6 Duties/Imperatives to *self-improvement* – This duty, we might say in a manner that is at odds with what psychology views as the process of self-actualization,

is focused primarily on understanding that our emerging and developing sense of "subject-hood" occurs *only* through our dealings with others and *Nature*, in and through *subjecting* ourselves to, in *subjugation* before, the imperatives of *Nature*. We become other *only* in our relationship with the other, in and through our immersion in the radiant light of the other's *face*. Self-improvement would importantly include a re-attuned attitude that would be grounded in the affective intuition, as related to *beneficence*, of the potential susceptibility that others have to instances of profuse and abject suffering, and a recognition of the *neediness* of the other.

7 *Duties/Imperatives to non-malevolence* – This is the imperative and duty to, as much as possible, ward off and resist doing evil, avoiding malicious behavior and actions that do unnecessary harm to others and *Nature*. As we have seen, this could be understood in the extreme on a grand scale as the avoidance of contributing to monumental ecological disaster, destruction, and devastation, or in the opposite direction, avoiding forms of harmful destruction of the environment on a smaller or micro-scale, whereby each one of us upholds the duty to avoid doing unnecessary harm to the natural world to which we are beholden.

In this chapter we have sought to carve out a middle road between the two forms of *ethical realism* Thomson associates with eco-phenomenology – recall that this view is grounded in the choice to embrace and incorporate the phenomenology of Husserl or Heidegger, as introduced earlier through our reading of Thomson. So, what our analysis represents is a unique "third way" to conceive of eco-ethics or the eco-phenomenological ethics of *Nature* that merges ethical intuitionism with phenomenology. Recall Thomson's claims regarding *Husserlian naturalistic ethical realism* and *Heideggerian transcendental ethical realism*, which are both forms of what Thomson identifies as "ethical perfectionism," which is a view to "ethical flourishing, broadly conceived, is best served by identifying, cultivating, and developing the distinctive traits or capacities of those entities deserving consideration."[137] Recall that *naturalistic ethical realism* contends that good and bad, or right and wrong, are determined on a reliance on the empirical "facts" of nature, and *transcendental ethical realism* looks to a transcendental source for ethical guidance – a transcendental signifier – that is gleaned when we are attuned to the environment. Since we have already shown the problems inherent within each of these views, we will not rehearse the critique here; however, we will pause briefly to examine how what we have developed within our philosophy of an ethics expressed through our *imperative duties to Nature* compares with and offers an alternative to each of these two versions of eco-phenomenological ethics.

To begin, it is undeniably the case that we embrace what Thomson has defined as *ethical perfectionism*, for in our analysis of both Ross and Lingis we have identified a multiplicity of distinctive traits and capacities of those things in *Nature* that not only deserve our consideration and attention, but beyond this, demand that we demonstrate a deep and sustained sense of ethical care for their valuable well-being, and this because they are worthy of respect because they *are* valuable in themselves. In both Ross and Lingis we encountered an ontological understanding of phenomenological *self-hood* and argued that the potential for *self-hood* and autonomy is dependent on the *other*, e.g., demonstrating a concern for the other's well-being, which might manifest in the situation where *Nature's* flourishing depends on its rehabilitation, which calls for subjecting ourselves to the duty to *reparations*, and the sustained effort to repair *Nature* becomes a crucial matter of interest, inspiring our respect and commanding us to uphold the prima facie duties to *beneficence* and to the avoidance of potentially eradicating *Nature* through our continued *malevolent* actions.

Our focus on Lingis, with our inclusion of the work of both Harman and Wheeler, allowed us to stress the notion of concern for the *other* beyond Levinas's *le visage* to include the *things of the world themselves*. In Lingis, we now understand that the focus on the potential for the *other*'s destruction, instantiates the attentiveness to the *other*'s sense of privation and *need*, with an awareness that the possibility always exists to arrest the other's authentic being or presencing when *fetishizing* what *should* be embraced as an *idol*. Lingis has provided us with wonderful descriptions of the great sequoias as *idols*, in moments when we are attuned to their imperatives, their unique lives and *needs*. Rolston, a major philosopher of environmental ethics, does indeed demonstrate certain *naturalist* tendencies when philosophizing his environmental ethics; indeed, in the extreme, as Toadvine observes, Rolston "seems to favor a teleological interpretation of nature drawn straight from Aristotle," in terms of constructing a "clockwork nature," within which the "gears of human activity can mesh."[138] However, despite this categorization, the manner in which he eloquently and insightfully speaks about an authentic eco-ethics and its focus on values, duties, and the human's role in *Nature* shares certain, albeit subtle, commonalities with the phenomenological interpretation we have developed. For example, Rolston intimates what it is like to undergo the transformation (*attunement*) from our anthropocentric view of *Nature*, which is haunted by the *Either/Or Fallacy*, i.e., believing that only humans are ethical subjects and nature is amoral; or that if nature has value, it can only be conferred upon it by a valuing subject, namely, humans; and in addition, that value is granted *only* because *Nature* serves to contribute to human flourishing. Below, attending to Rolston's description we can map on the ethical insights gleaned from Lingis's phenomenology, and follow the attuned movement or transition from the uniformed and heedless *fetishizing* of *Nature* to the re-attuned *ethical* attitude instantiating respect and reverence for *Nature* as an *originary* imperative:

> In Yosemite National Park for almost a century humans entertained themselves by driving through a tunnel cut in a giant sequoia. Two decades ago the Wawano tree, weakened by the cut, blew down in a storm. People said: Cut us another drive-through sequoia. The Yosemite environmental ethic, deepening over the years, said no! You ought not to mutilate majestic sequoias for amusement. Respect their life![139]

As stated, when pursuing eco-phenomenology, we look to thinkers who did not develop an eco-ethics, yet we incorporate their work in the effort to contribute to developing a coherent and legitimate, albeit tentative and speculative, view of phenomenological ethics related to *Nature*. The prima facie *imperative duty ethics* presented cannot, as in *naturalistic ethical realism*, be traced to empirical "facts" about *Nature*, and this we have understood as the radicalization of our inability to have "knowledge" of the most essential aspects of the things we encounter, as we find in Lingis's phenomenology and Harman's OOO, as well as in the prima facie *ethics* of Ross, which as we have seen does not rely on "facts" or formable knowledge, but rather on our "apprehension" or intuition to glean that which is right and good. Interestingly, Rolston also recognizes that a sound ethics of the environment, in a way that relates directly to the arguments we have formulated, must move beyond "science" and have "everything to do with beliefs about duty,"[140] i.e., finding "values in a duty to the natural world."[141] This indicates that when the world or *Nature* shows itself to us, as intrinsically valuable, it at once "informs [us] the way it *ought* to be"[142] and the ways that we *ought* to be based on the enlightened understanding that we are beholden as subject to *Nature*'s imperatives. We have sought to avoid slipping into the trap of anthropocentricism, which again, as Rolston states, is a bane facing many *naturalistic* views of environmental ethics, for an authentic ethics of *Nature* should "seek an appropriate respect for life," with the understanding that "respect for human life is only a subset of respect for all life,"[143] and the valuing subject "in an otherwise valueless world is an insufficient premise for experienced conclusions of those who respect all life,"[144] because there is "no reason to say that all value is an irreducible emergent at the human (or upper animal) level."[145] We have also tried to present a view of ethics that, unlike *transcendental ethical realism*, gives us distinct values that can be drawn out from our encounters with *Nature*, which can be identified, expressed, and practiced. Thus, we have embraced ethical meanings that can be concretized, where values are understandable and transmittable, and so have attempted to retain a sense of "practicability" and "publicity," while at once recognizing that a rigid, indelible sense of "universalizability" eludes us, in the sense that we cannot provide what amounts to "valid" and "sound" arguments for our claims; but as consistent with our position, in the *normative* realm the quest for such a high degree of "certainty" is highly misplaced. And yet, let us consider the following: Since our project focused on the ontological and the existential, the universalizability of our prima facie *imperative duty* ethics might already live as a presupposition, i.e., *normativity* is ineluctably bound up with an *ontology* of the human and *Nature*. Although not embracing a strict *transcendental ethical realism* as philosophized by Thomson, what we have presented certainly holds the potential to inspire a form of ethical "*humanistic* perfectionism" – but this is *not* to be equated with yet another manifestation of *secular humanism* – i.e., we have philosophized a *perfectionism* that contributes to because it "emphasizes the cultivation of the *distinctive* traits of *Dasein*."[146]

Notes

1 Charles Brown and Ted Toadvine, "Eco-Phenomenology: An Introduction," in *Eco-Phenomenology: Back to the Earth Itself*, eds. Charles Brown and Ted Toadvine (Albany: SUNY Press, 2003), xix.
2 Charles Brown, "The Real and the Good," in *Eco-Phenomenology: Back to the Earth Itself*, eds. Charles Brown and Ted Toadvine (Albany: SUNY Press, 2003), 9.
3 Ibid.
4 Ibid., 10.
5 Peter Angeles, *Philosophy* (New York: Harper-Collins, 1992), 206.
6 John Nolt, "The Move from *Good* to *Ought* in Environmental Ethics," *Environmental Ethics*, 28, no. 1 (2006): 358.
7 Ibid.
8 Ibid., 360.
9 Brown and Toadvine, "Eco-Phenomenology: An Introduction," xii.
10 Brown, "The Real and the Good," 11.
11 Iain Thomson, "Ontology and Ethics at the Intersection of Phenomenology and Environmental Philosophy," *Inquiry*, 47, no. 1 (2004): 382.
12 Brown, "The Real and the Good," 13.
13 Ibid.
14 Ibid., 15.
15 David Detmer, *Phenomenology Explained: From Experience to Insight* (Chicago: Open Court, 2013), 15.
16 Ibid.
17 Thomson, "Ontology and Ethics," 386.
18 Ibid., 389.
19 Ibid., 10.
20 Ibid.
21 W. D. Ross, *The Right and the Good* (New York: Oxford University Press, 1961), 17.
22 Ibid., 25.
23 Ibid., 26.
24 Ibid., 27.
25 Ibid., 30.
26 Aristotle, *Topics Books I and VIII*, trans. Robin Smith (New York: Clarendon Press, 1997).
27 Ross, *The Right and the Good*, 29–30.
28 Ibid., 28.
29 Graham Harman, *Object-Oriented Ontology: A New Theory of Everything* (London: Penguin, 2018), 35.
30 Ibid., 36.
31 David Best, "Feeling and Reason in the Arts: The Rationality of Feeling," in *The Symbolic Order*, ed. Peter Abbs (London: Falmer Press, 1989), 70.
32 Ibid., 70–71.
33 Ross, *The Right and the Good*, 24.
34 Ibid., 23.
35 Ibid., 29.
36 Ibid., 30.
37 Ibid., 31.
38 Don Marietta, "Back to Earth with Reflection and Ecology," in *Eco-Phenomenology: Back to the Earth Itself*, eds. Charles Brown and Ted Toadvine (Albany: SUNY Press, 2003), 131.
39 Ibid., 131–132.
40 Louis Pojman and James Fieser, *Ethics: Discovering Right and Wrong* (Belmont, CA: Wadsworth Press, 2009), 124.

41 Samuel Pufendorf, *On the Duty of Man and Citizen According to Natural Law* (Cambridge: Cambridge University Press, 1991), 57.
42 Ross, *The Right and the Good*, 25.
43 Hilary Putnam, *The Many Faces of Realism* (LaSalle: Open Court, 1987), 85.
44 Holmes Rolston III, "Challenges in Environmental Ethics," in *Environmental Philosophy: From Animal Rights to Radical Ecology*, eds. Michael Zimmerman, et al. (Englewood Cliffs: Prentice Hall, 1993), 154.
45 Rudolf Arnheim, "The Double-Edged Mind: Intuition and the Intellect," in *The Symbolic Order*, ed. Peter Abbs (London: Falmer Press, 1989), 107.
46 Ibid., 108.
47 Ibid., 102.
48 Ibid., 107.
49 Louis Arnaud Reid, "The Arts Within a Plural Concept of Knowledge," in *The Symbolic Order*, ed. Peter Abbs (London: Falmer Press, 1989), 18.
50 Ibid.
51 Best, "Feeling and Reason in the Arts," 70.
52 Ibid., 78.
53 Sonia Kaminska, "The Alleged Activity of the Active Intellect: A Wild Goose Chase or a Puzzle to be Solved? *Zagadnienia Filozoficzne w Nauce*, 54, no. 1, 82.
54 Michael Bonnett, *Moral Education and Environmental Concern* (New York: Routledge, 2013), 19.
55 Ibid., 20.
56 Michael Bonnett, "Environmental Education and the Issue of Nature," *Journal of Curriculum Studies*, 39, no. 6 (2007): 714.
57 Ibid., 716.
58 Ibid.
59 Holmes Rolston III, *Environmental Ethics: Duties to and Values in the Natural World* (Philadelphia: Temple University Press, 1981), 230.
60 Ibid., 231.
61 Michael Bonnett, *Retrieving Nature: Education for a Post-Humanistic Age* (United Kingdom: Wiley-Blackwell, 2004), 81.
62 Nolt, "The Move from *Good* to *Ought*," 367–373.
63 Bonnett, *Retrieving Nature*, 88.
64 Ibid.
65 Randolph C. Wheeler, "Moral Subjectivity and Imperative Subjection" (2006): 3, www.academia.edu/7343439/.
66 Ibid., 4.
67 Ibid., 3.
68 Ibid.
69 Ibid., 4.
70 Ibid.
71 Nolt, "The Move from *Good* to *Ought*," 364.
72 Michael Bonnett, "The Powers that Be: Environmental Education and the Transcendent, *Policy Futures in Education*, 19, no. 2 (2015): 6.
73 Ibid., 7.
74 Ibid., 6–7.
75 Ibid., 15.
76 Alphonso Lingis, *The Imperative* (Bloomington: Indiana University Press, 1998), 172.
77 Wheeler, "Moral Subjectivity," 16.
78 Ibid.
79 Lingis, *The Imperative*, 172.
80 Ibid.
81 Ibid., 173.
82 Ibid., 172.

83 Ibid., 184.
84 Bonnett, "The Powers that Be," 16.
85 Randolph C. Wheeler, "Imperative Innovations," in *Passion in Philosophy: Essays in Honor of Alphonso Lingis*, ed. Randolph C. Wheeler (New York: Lexington Books, 2017), 127.
86 Wheeler, "Moral Subjectivity," 16.
87 Ibid., 17.
88 Ibid., 3.
89 Lingis, *The Imperative*, 120–121.
90 Wheeler, "Moral Subjectivity," 3.
91 Alphonso Lingis, *Foreign Bodies*, (London: Routledge, 1994), 3.
92 Ibid., 4.
93 Ibid., 11.
94 Liddell and Scott, *A Lexicon: Liddell & Scott's Greek-English Lexicon* (Mansfield Centre: Martino Publishing, 2015), 144.
95 Lingis, *Foreign Bodies*, 11.
96 Lingis, *The Imperative*, 53.
97 Lingis, *Foreign Bodies*, 13.
98 Lingis, *The Imperative*, 181.
99 Ibid., 180.
100 Ibid., 172.
101 Ibid., 220.
102 Ibid., 221.
103 Emmanuel Levinas, *Totality and Infinity*, trans. Alphonso Lingis (The Hague: Martinus Nijhoff, 1961), 211/193.
104 Wheeler, "Imperative Innovations," 125.
105 Ibid.
106 Wheeler, "Moral Subjectivity," 15.
107 Ibid.
108 Graham Harman, *Towards Speculative Realism* (New York: Zone Books, 2009), 18.
109 Ibid., 20.
110 Ibid., 18.
111 Ibid., 17.
112 Alphonso Lingis, *The Community of Those Who Have Nothing in Common* (Bloomington: Indiana University Press, 1994), 45.
113 Ibid., 64.
114 Wheeler, "Imperative Innovations," 122.
115 Lingis, *The Imperative*, 220.
116 Ibid., 174.
117 Ibid., 172.
118 Ibid., 174.
119 Ross, *The Right and the Good*, 155.
120 Ibid.
121 Immanuel Kant, *The Metaphysics of Morals*, trans. Mary Gregor (Cambridge: Cambridge University Press, 1998), 160.
122 Ibid.
123 Ibid.
124 Ross, *The Right and the Good*, 172.
125 Ibid., 173.
126 Lingis, *The Imperative*, 63.
127 Ibid., 64.
128 Ibid., 68.
129 Ibid., 9.
130 Harman, *Towards Speculative Realism*, 17.

131 Ross, *The Right and the Good*, 22.
132 Lingis, *The Imperative*, 172.
133 Ibid.
134 Lingis, *The Community of Those Who Have Nothing in Common*, 66.
135 Ross, *The Right and the Good*, 27.
136 Lingis, *The Imperative*, 63.
137 Thomson, "Ontology and Ethics," 386.
138 Rolston, "Challenges in Environmental Ethics," 141.
139 Ibid.
140 Ibid., 156.
141 Ibid., 136.
142 Ibid., 156.
143 Ibid., 136.
144 Ibid., 154.
145 Ibid.
146 Thomson, "Ontology and Ethics," 386.

5 The phenomenology of human-and-*Nature* involvements

The ethical call of Nature's reticent imperatives

Immersed, embodied, and emplaced intentionality

A reconceived notion of the human-and-Nature relationship

Reconceiving correlationism as expressive of our intimate relationship with Nature

Recall that Harman levels a damning critique against traditional metaphysics, with its "philosophy of access" – which is "concerned only with our access to the world and never with the world in its own right,"[1] – locked into a hardened and limited version of *correlationism*, which we have seen is a concern with the interaction of things/objects in the world in terms of the tensions that exist between objects and their sensual qualities, and not exclusively with "the one *type* of relation that becomes central to all philosophy – the duel between human and world."[2] And, as we have seen, for Harman, "sentient creatures are just a more advanced case of a universal drama between any two things."[3] However, for our eco-phenomenological project, we are concerned with a reconceived vision of *correlationism*, the relationship between human/thought-and-world/things, with the understanding that there is a world of things and objects whose existence, interactions, and relationships do not depend on the correlation between human thought and world, and so, importantly, are independent of human involvement. We must keep in view the notion of *Nature*'s intrinsic value developed in the last chapter, for as related to *correlationism*, Bonnett reminds us that, and this harkens to our analysis of Harman's OOO, although values might only "occur – come to light – in the space of human consciousness (or its equivalent), they are not entirely anthropogenic, not purely a product of human estimation and grading of things where the metric is human-authored."[4] For example, when in the presence of beauty, it enraptures us, and in the thrall of its attunement, we "can be *given* a metric rather than apply or impose one of our own device."[5] With that stated, we are still concerned with *how* we interact with the *things of the world*, the things, objects, and entities of *Nature*, and so when turning to Bonnett's phenomenological view of *Nature*, we come to understand the following: Just because we can draw out a legitimate distinction between

> what is experienced as human and what is experienced as nature, this does not make the view dualistic in the sense of supposing that nature is authentically understood as an object for an independent subjectivity, denying that we exist *in* nature, or that consciousness is in an important sense part of nature.[6]

It simply indicates that human consciousness, as embodied-intentionality, is of a unique makeup, it is the context within which "things show up and they are encountered, are understood, and become intelligible *as* things (mysterious and inhering in a world),"[7] as existing in such a way that our intentionality is dependent on the world of things in which we are immersed, i.e., dependent on *how* we are with and the manner in which we *dwell* within *Nature*. As is consistent with eco-phenomenology, seeking to explore and deepen our understanding of the relationship between humans and *Nature* will require the typical phenomenological move beyond the naturalistic and biological world of abstraction.

We now know that in Lingis's phenomenological view of "corporeal intentionality" and the notion of imperatives in the *things themselves*, there is a focus on the reconceived correlation between human-and-world, and so we also encounter a radically reconceived understanding of *correlationism*, which importantly, related to eco-phenomenology, transcends the deleterious attunement of the anthropocentrism at the core of the *metaphysics of presence*. These issues emerge from his early analysis of both Husserl and Merleau-Ponty, as well as his later writings, which provides us with the phenomenological understanding that we are immersed in the world and intimately connected to it through our perceptual, cognitive, and ethical responses, all of which might be gathered under the phenomenon of "immersed-intentionality" as it is inseparable from and dependent on our responses to the imperatives emanating from the *faces* of others and *Nature*. In this final chapter, we first explore the eco-phenomenological issues of intentionality through a reconceived understanding of *correlationism* that transcends the anthropocentric dualism (subject(ive)-object(ive) divide) haunting traditional *productionist* metaphysics and ecological philosophies emerging from this view. Here, we consider the engaged context of human-and-*Nature* interaction in terms of what Bonnett describes as a realm of "emplacement," within which we are immersed in *Nature*'s unfolding as responsible denizens, dwelling in a living context highlighted by perceptual and intuitional receptivity, intimately and ineluctably set within the cycle of birth, growth, degeneration, and death, i.e., *existentially* stretched out temporally inhabiting a *space* between *finitude* and *transcendence*. Second, we deepen this initial analysis by exploring the ineluctable ontological relationship that both the human and *Nature* have with *temporality*, *space*, and *transcendence* as related to human *subject-hood* and the ontological condition of *finitude*. We introduce the concept of the human's relationship with *Nature* and show that it is grounded in the ontological phenomenon of *sonority*, wherein the other's *presencing* and recession into concealment and silence occurs within an ever-renewed "dialogue" structured through the process of *address and rejoinder* between the human-and-*Nature*. Third, building on and deepening the phenomenological understanding of *sonority*, we return to the notion of imperatives in *Nature* by exploring the manifold ways in which the natural world "speaks" to us. Pursuing this line of thought, which we relate to the ontological-existential theme of *communality*, we analyze the issue of *sonority* and our resolute and anticipatory attitude toward the *normative openness* required to listen for and ultimately respond appropriately to the call or address of *Nature*, which "sounds," from out of the depths of its *affective materiality* as a living force, through whispers, primordial bestial wails, animal outcries, stirrings, and a mode of communication that is at times akin to a deep ontological "murmur," a low distinct and reverberating hum, which often communicates through muffled, half-suppressed "resonations" producing an emotional response that connects us with and relates us to an entire community of life. This analysis incorporates elements from the foregoing chapters in the attempt to provide a unique and thorough understanding of the reconceived and renewed eco-phenomenological vision of the human-and-*Nature* relationship.

Finite transcendence

Recall the four main ontological "themes," or Sense (*Sinn*)-giving-making structures, and how we noted that these themes have been dealt with extensively, in and through a variety of unique ways by incorporating differing methods and approaches, e.g., in the *eidetic* phenomenology of Husserl, the ontological phenomenological-hermeneutics of Heidegger, the *Existenz* philosophy of Jaspers, and the existential-phenomenology of Sartre. We identified those *themes* as follows: *Lived space* (*spatiality*), *lived body* (*corporeality*), *lived time* (*temporality*), and *lived human relations* (*relationality* or *communality*).[8] Since our concern is eco-phenomenological, it is necessary to reassess these ontological *themes* in order to modify them and relate them specifically to our analysis. As introduced, what Lingis refers to as "corporeal intentionality" and we have referenced as "immersed-intentionality" Bonnett refers to as "emplaced transcendence," and we must first provide a definition of transcendence prior to exploring ontological themes we draw from our exploration of the human-and-*Nature* relationship.

The term transcendence as we understand it and employ it is not a synonym for the "transcendent," as that which stands beyond the human condition, as in the transcendental signifier discussed in relation to *transcendental ethical realism*. It is also the case that transcendence is neither a *vertical* phenomenon, as in the basic or general religious sense of transcendence, nor is it a *horizontal* phenomenon, as would be consistent with Gray's understanding of *secular* "eschatological" (end-of-history) philosophies as first represented in Hegel and Marx, a view which emerges from and is traceable to the theological (religious) view of eschatology found in, to name but one source, the canonical *Gospel of Mark*.[9] Both these views stand in contrast to our eco-phenomenological view, and they adopt a linear view of time as related to the everyday understanding and the method of historical analysis found in *narratology*. If we turn to Heidegger, and look past the critical issue of the latent subjectivity in the *fundamental ontology* of Dasein as is common in literature focused on the "Turn" in Heidegger's thinking after *Being in Time*, we encounter a definition of transcendence that speaks directly to our concerns. Heidegger informs us that *Überstieg*, in a straightforward manner, means "surpassing" or "transcendence," expressive of a relation that "passes 'from' something 'to' something."[10] Here, we might understand transcendence as a process of change or evolution where, in our case, we *become other to ourselves in the face of the other*. In succinct terms, transcendence instantiates human flourishing, a form of self-overcoming where we deepen our sense of both understanding about and concerned connection to the world and others. However, in relation to *finitude*, the radical limitations giving ontological structure to the human condition, this is properly a form of *finite transcendence*, linked inextricably with the radical limitations of human knowledge.

In stating that transcendence might be understood as *becoming other in the face of the other*, we recall that "other" refers not only to humans, but also to things, objects, entities, and *Nature*. Bonnett relates transcendence to *Nature* in the following manner: Our immersed-intentional states are always grounded in our participating in *Nature*'s transcendent aspects as well as the human's potential for transcendence as related to *self-hood*, for when engaging *Nature*:

1 There is a sense of mystery and otherness that permeates and provides organization to our experience – things neither merely come to presence before us nor do they reveal themselves in their essential or complete unconcealment;
2 The mystery in things is experienced as *normative*, and as related to what we have shown in Chapter 4, it places demands on us and opens the potential for what constitute fitting responses on our part to *Nature*;
3 In the midst of the occurrence of revealing and engaging things, "in their becoming and in their passing ... we are brought back to the domain of elemental powers that condition our lives and bind us to the cosmos."[11]

Here, to return to our understanding of *Überstieg*, self-overcoming and the potential for *self-hood*, which is the *finite* process of "transcending," as we also encounter in Lingis, is ineluctably linked to the ontological or phenomenological sense of *self-hood*, which is wholly contingent on our interaction and relationship with *Nature*. As Wheeler rightly observes, Lingis presents a view of *self-hood* that outstrips the "subject" common to the anthropocentric humanist tradition. Our immersed-intentionality or our "emplacement" in *Nature* is highlighted by a *mutual* sense of *anticipation*, in that we are set within a world or "milieu of prompts and claims that occur in particular places and that are always partly familiar,"[12] and within such places – experienced in terms of ontological *spatiality*, *temporality*, and *communality* – we are always open in an *anticipatory* manner to offering unique rejoinders to the call of *Nature*. In this mode or attitude of *anticipation*, we stand outside ourselves, in the sense that we instantiate a potential or imminent *ek-stasis*, as related to transcendence and the manner of being open to new possibilities that are on the approach from out of the indeterminate future, and which contribute in non-determinative and unforeseen ways to our developing *subject-hood*. From the perspective of an existential understanding, we are always already other (*ek-static*) than what we embody in any given and present moment – as we *stand-out* from and *beyond* ourselves – and "we are enlivened by this capacity to be beyond ourselves that enables felt encounters with what is other,"[13] which importantly includes our potential to be attuned to modes of world-disclosure other than those embraced by and giving structure to the overarching "metaphysical" CF. But, as we learn, Bonnett, in a manner similar but not identical to Lingis, brings our attention to the "dialogue" that we instantiate with *Nature*, and makes the case for *Nature*'s participation in this "dialogue," which as stated, is essential for the continued development and evolution of our *self-hood* as grounded in a relationship that is essentially *normative*.

Immersed-intentionality as human emplacement in Nature

Recall our analysis of Harman's OOO and the reductive view of objects, which when applied to *Nature* might be expressive of a *materialist* view, built on at least two patently false assumptions: "*everything that exists must be basic and simple*" and "*everything that exists must be physical*."[14] This, as we have seen, in eco-phenomenology, represents a mechanistic scientific view of *Nature*, and, as Toadvine contends in relation to our analysis of eco-ethics in Chapter 4, this *naturalistic* conception "reduces nature to axiologically neutral raw material."[15] We also saw that even in views shunning materialist conceptions of *Nature* and embracing an eco-philosophical view, such as the *Deep Ecology* movement, the propensity exists to continue to harbor the underlying scientific and hence *naturalistic* tendencies of materialism, e.g., Naess's *relational total-field* image of *Nature* retains the biological understanding of human-and-nature, which might be restated in terms of the enhanced view of *human-in-environment*, or in more extreme terms, *organism-in-environment*. As Bonnett rightly concludes, and as has been the running theme of our research, the aforementioned views shape our understanding of our *place* in *Nature* and can have crucial implications for the ways we view ourselves and our role as responsible ethical agents, i.e., not only the manner in which we behave toward each other, but also the way we comport ourselves and hence take a stance within and toward *Nature*. Our *placement* or *emplacement* in *Nature* should not be conceived as our physical placement within a grand abstract system, but rather in a more intimate manner, in terms of "our felt sense of place – and potentially of nature's presence – in the intimacy of our being in the particular locales that constitute the life-world of our daily experience."[16] This *emplacement* as immersed-intentionality, as Bonnett suggests, is no longer an experience either given order by theoretical conceptual schemas or driven by instrumental purposes, but rather it is an "embodied" experience that puts us in touch with and sets us within the "play of elemental powers" structuring and conditioning the spaces of *Nature*, the unfolding of primordial phenomena within which we participate such as "birth and death, growth and decay, sound and silence, lightening and darkening, movement and stillness";[17] and in relation to the inherent value of *Nature*, we are open to experience the intuitive sense of the ethical rightness and wrongness of certain behaviors that instantiate our responses to *Nature*. When considering the unfolding of these elemental forces and our place among them and relation to them, we should recall the following important issues: Harman's view of the "tragic burden" of *finitude* and Lingis's claim that it is necessary that we care for all things in the exigent response to their potential for incurring harm. Ultimately, these issues relate to what we might call the overarching concern of the eco-phenomenological project, namely, "the very idea of the earth's mortality" that explains "the urgency in the call for an ethical response that the experience, direct or otherwise, of the growing ecological disaster solicits."[18]

When engaging *Nature*, to revisit Lingis's notion of "corporeal intentionality," our comportment neither occurs as a purely *eidetic* and hence internal and "subjective" phenomenon, nor can it be expressed through abstract models, and Wood, in agreement, states in no uncertain terms: "[I]ntentionality is firmly lodged within my bodily existence, within the natural world."[19] Bonnett echoes this view, asserting that our engagement with and immersion in *Nature* is a fully embodied, multi-sensory experience of both feeling and cognition, an experience that holds the potential to shape and engender a generative form of *selfhood*. For to discover meaning or engage in the hermeneutic process of heuristic meaning-making highlighting our relationship with *Nature* is never reducible to a "subjective layer projected on brute reality, having no internal connection or reference to that reality and therefore no special validity – as if it all comes from us, is fully authored, sired, by us."[20] The external world we have philosophized is neither a "Cartesian inference" nor a "hypothesis," but is rather "what is closest to us; phenomenologically, we are embedded in it and it lies within us."[21] Indeed, because we live in the presence of things as *ek-static* beings, we instantiate a "reflexivity necessary for self-awareness and the sensing of our emplacement in the world – a world of infinitely extended and interconnected significances made manifest through the presencing of things," which inspires an attunement where we respond to what is "present and underway: a dimension of being yet to be fulfilled,"[22] a dimension rife with potential and possibility, requiring an anticipatory attentiveness and an attuned listening on our part – as we later address in relation to Lingis's phenomenology. And yet, we must be clear, in relation to a theme we have stressed throughout, things of the world, of *Nature*, proceed in their own inexplicable and ineffable ways, for theirs is an existence that essentially remains, proximally and for the most part, mysterious to us, for *Nature*'s truths are *excessive* and transcend what is given to us directly in experience, and this we call the supreme mystery of *Nature*, which has the potential "for taking us beyond ourselves, which itself is a requirement for moral relationships."[23] This relates to our analysis in Chapter 4, indicating that things presence before us in ways that communicate *how* they *should* be treated, how they *ought to be* appropriated, and we intuit what counts as the fulfillment of their unique needs; and by at once participating in and sustaining this *dialogic* process of *need*-and-*response* or *call*-and-*rejoinder*, "we, too, are sustained by our participation" in a way of dwelling in *Nature* that Bonnett labels "place-making."[24]

Our relation to and with *Nature*, as a *place-making* endeavor, unfolds through an immediate and sensual interaction with things that are *emplaced*, i.e., things unfold as they share and contribute to establishing a *common place*, and this we might conceive as "rooted-ness," which means that we are immersed in the unfolding of the elemental and ontological primordial forces of *Nature*, and this Bonnett labels "autochthony." We note that for the Greeks, *autochthenes* means that human beings are *of* and *from* the earth, and hence, not merely related to, but more than this, inseparable from the earth or *Nature* (*physis*), i.e., this Greek term might be thought of as expressing a view of our ontological immersion in the primordial, non-exclusive native (human) soil *within* which we *all* participate and *from* out of which we all arise and return to in death. This intimate experience of living and dwelling instantiates a realm of involvement, inspiring the solicitous caring that attunes us to "receive what is gifted and gift what is received," which leads to us formulating and offering fitting responses to *Nature*, and it is through this process of responding to *Nature* that the *anticipatory*, *ek-static*, and *transcendent* aspects of life as comprising and contributing to our *self-hood* manifest as authentic possibilities, and in this context of "dialogue" between human-and-*Nature*, the *metaphysics of presence* has no place to gain a foothold. This notion of "emplaced transcendence" and immersed-intentionality developed in the preceding paragraphs, as stated, occurs within a *space* or *place of dwelling*, and we now move to consider the makeup of this *place* of mutual interaction as related to the phenomenological-ontological themes of *spatiality*, as philosophized in the eco-phenomenology of Bonnett, and *temporality*, as expressed in the eco-phenomenology of Wood. As stated, our emplacement in the context of *Nature*'s unfolding determines *who* we are and, in a crucial and significant manner, shapes our Being-in-the-world, and in this ontological context of *emplaced-attunement* our pre-reflective experience of "space" cannot be reduced to a geometric understanding and our "lived" enactment of "time" is radically other than our everyday experience of linear, calculable, and measurable "clock-time."

Nature and ontological-existential "themes": spatiality, temporality, and communality

Heidegger's phenomenology, or the *fundamental ontology* of *Dasein*, presents us with a sustained discussion of "ontological space," or *spatiality*, highlighted by two revolutionary notions, *de-distancing* and *directionality*, and with these ideas Heidegger reveals an intricate phenomenology of "space" and "dwelling." Heidegger, in a trend common to phenomenology, marks off ontological space from Cartesian space, or geometric space, which is located in a view of the *res extensa*, or more properly, the world of extended matter or space as a three-dimensional phenomenon, i.e., the type of space we *fill*, the type of space as "distance" that we *traverse* or *cover*. However, to reductively conceive space only in terms of remoteness or closeness, as *distance*, or space as a medium within which all things are set and move, misses and ignores the phenomenological or ontological understanding of the authentic "lived experience" of space, i.e., simply bringing things closer to us in geometric proximity is not the equivalent or experience of ontological *spatiality*, which instantiates the ontological proximity of the world and things to our essential Being. Heidegger is concerned specifically with Dasein and its Being-in-the-world, whereas our concern is with the relationship between the human and *Nature*, so with these brief remarks we return to the eco-phenomenology of Bonnett who presents many interesting ideas regarding space – as *spatiality* – in our relation to the manner in which the human being and the things of *Nature* are set within, nested in, as it were, a space or place of dwelling that houses, shelters, and facilitates a *way of dwelling*, which is irreducible to the understanding of space that can be measured and determined by Cartesian coordinates.

When considering Bonnett's view of the context of existential *spatiality*, the space of *Nature* is composed of "distinct neighborhoods characterized by changing local ambiences, events and significances,"[25] which claim us and inspire in us views, feelings, and perceptions that determine the manner in which we are *present* in that *space*, attuning us to behave in certain ways that demonstrate respect for the *dwelling space* and its inhabitants. Proximity in such a context is experienced as the closeness of *Nature* to our Being and our Being to *Nature*, manifesting through a process wherein each of us that inhabits the place participates in the reciprocal relations that constitute the place, i.e., each of us in unique ways contributes to *place-building*, working to establish and sustain the dwelling space we share. Much like Bonnett and Lingis – for recall the perceptual and axiological organizational powers of *Nature*'s imperatives that demand our response and respect – Wood also locates a sense of integrity in *Nature*'s organization and its unfolding through what we have already labeled *elemental* or *primordial forces* giving rise to and directing the natural "processes, births, deaths and catastrophes as well as the continuities"; and in line with our view, this process of organization and its unfolding can be understood in a biological manner, or as Wood asserts, in terms of epistemology. However, more importantly for eco-phenomenology, these "concerns are of an ontological nature."[26] In Wood's eco-phenomenology we also encounter a notion of *spatiality* in his elucidation of what he terms the "boundaries of thinghood," for when we consider the unity of things and the integrity of their organization, immanent in their relationship is the active establishment and maintenance of the boundaries marking out and separating off what resides outside of them. These boundary zones establish an existential sense of what is inside and what must necessarily remain outside if the integrity of *spatiality* or the dwelling space is to be maintained; and here, too, there is the undeniable ambiance of *normativity*. Boundaries, however, change and evolve, for as Wood contends, they are "sites of negotiation, of transformation, of sustenance, and of protection," and they are "products of the very processes that maintain them."[27] Boundaries are established sites of what Wood terms "linima," or what we might understand in terms of *thresholds* for either the facilitation or destruction of the boundaries of *spatiality*. When these *thresholds* or ontological *limits* are recognized and respected, this facilitates the appropriate boundaries of *spatiality*, but when traversed and overstepped, this leads to the destruction of the self-same boundaries. *Thresholds*, we note, are also given form and limitation by *temporality*, or what we understand as "lived time," for as Wood emphasizes, eco-phenomenology must be concerned, in addition to *spatiality*, with the "fabric of time and with the events that occur at boundaries."[28]

In line with Wood, Bonnett observes that just as there is a process of *place-making* and the sustaining of that place, there are also ways in which we can destroy existential *spatiality*. For example, to concretize this notion, to transplant an indigenous tree from its "en-forested" dwelling to the center of a shopping mall is to destroy the existential *spatiality*, the context of its intimate and authentic dwelling, for it is not simply the case that "some objects have been removed from the place, it is rather that havoc has been wrought on the place itself,"[29] because the ontological interplay of things has been disrupted. Not only does this influence the unfolding of the interrelationships comprising *Nature*, but it also, if we keep in mind what we have said regarding the bond between our interaction with *Nature* and human *self-hood*, influences the course and direction of our development, which is a matter for our potential to enact *finite transcendence*, and which, again, cannot be thought of as apart from our *emplaced* dwelling in *Nature*. All of the phenomenological issues related to *self-hood*, such as the *anticipatory* listening to and for the other, the *ek-static* potential to stand-out from our current and present modes of dwelling, and the issue of *transcendence*, which is essential for the *mutual anticipation* of encountering the other, are disrupted by actions that compromise and degrade ontological *spatiality*. We note that in appropriating the above example from Bonnett, we do not mean to trivialize the negative effect that such *de-placement* or *dis-placement* has; but the removal of a single tree from its emplaced dwelling does not inflict the same magnitude of *ontological damage* to the dwelling context (human-and-*Nature*) as the destruction brought to a large portion of the Alaskan coastline – the "place" or "space" of Prince William Sound – with all of its mutual dependencies and interconnected relations, wrought by the massive oil spill (EVD) already discussed in Chapter 2. Ultimately, what we sacrifice when destroying the context of emplacement, the existential instantiation of *spatiality*, is the integrity of *Nature*, which as we have seen, is expressive in an essential manner of its intrinsic worth and value. We can also imagine, in an all too real and relatable way, the interruption or destruction of ontological *spatiality* when we consider the phenomenon that we shall refer to as the invasion or introduction of the *foreign* or *alien* into the *intimate*-and-*familiar*. Indeed, Lingis asserts that when we open our home (space) in way that creates "the zone of the alien," we then inhabit "the region where the order of the home enters into decomposition,"[30] for we must be clear that what is *alien* is not the equivalent of what is *other* as we have discussed and embraced the latter.

This notion can be understood by considering the issue of mass pollution occurring not only to our local waterways, streams and rivers, but also to the oceans of the world. To illustrate, we turn to a scenario from Rolston, noting that although his concern is not with existential *spatiality*, but rather the immorality of contributing to pollution and allowing it to destroy wildlife habitats, it is possible, through an analysis of his description, to wrest from concealment the issues of ontological *spatiality* and existential *boundaries* thus presented. Rolston informs us about Atlantic fishermen of the coastal estuaries who discard their trash, which includes all of their empty beer bottles, by throwing it overboard. Despite the seemingly inconsequential nature of this practice, born perhaps of an unconcerned ignorance, it radically disrupts the dwelling space and the "emplaced" nature of marine life within their familiar "neighborhoods." For on the ocean floor, "small crabs, attracted by the residual beer, make their way inside the bottles and become trapped, unable to get enough foothold on the slick glass neck to work their way out. They slowly starve."[31] The dead carcass of the entrapped crab then serves as bait for the next crab that will suffer the identical fate. This instantiates what we have labeled the *alien* within the *familiar*, and in this instance *boundaries* are transgressed and defiled. We also encounter this phenomenon when considering the ever-widening blight of pollution that results from discarded plastic products scattered across beaches, clogging waterways, and eventually making their way into the ocean to wreak havoc on marine life and the local marine ecosystem. Again, this is clearly an instance where the existential boundaries established by *Nature* are crossed and eventually obliterated. For recall Wood's philosophizing of boundaries and his focus on "boundary maintenance," where acknowledging and respecting boundaries facilitates cooperation and dependency, and recall too that the converse is also the case, i.e., the disruption of boundaries leads to "rupture, catastrophe, and [unwanted and unhealthy] transformation."[32] Consider that small pieces of plastic remains are often eaten by a wide variety of sea creatures, such as fish, turtles, and seabirds, that come in contact with the plastic on beaches and in the oceans, often resulting in their death. Aquatic wildlife often get entangled in plastic debris, leading to injury and potential death, for we must be aware that plastic does not completely biodegrade – it accumulates in the ocean, polluting natural habitats, and when it does begin to break down it produces toxic particles called "microplastics," which enter the food chain, impacting human health when fish or other animals that have been poisoned by the microplastic particles are consumed. Accordingly, the "vast swirls of plastic rubbish visible on the sea surface – horrifying as they are – represent just the tip of the iceberg,"[33] for the problem inevitably, if not promptly addressed, makes its way below the surface to disrupt the life cycles of large numbers of marine denizens, and as if this were not bad enough, the "plastic crisis" in the ocean also contributes to perpetuating Global Warming.[34] Here, although what we have described is undeniably a biological and empirical problem, it also represents an *existential crisis* in the true sense of the concept, for it is the case that in a more *originary* manner, such problems are initially traceable to the loss of or disruption to the primordial sense of *spatiality* that is instantiated within the lives of communities of marine animals in their natural dwelling places, all of which are given structure by the *boundaries* that help perpetuate a thriving life and existence.

As referenced in Chapter 1, as related to Wood's eco-phenomenology, to return to a pre-reflective realm of dwelling is not to open up a space into what might, in an abstruse sense, be understood in terms of a "pure presence"; rather, as we have seen, it is to "return to a world in which the relation between present experience and the complexity of what is being experienced has always been deep and stratified."[35] And what is deep and stratified can be understood as inherently instantiating a structure and order that is related to the ontological *themes* that provide structure for the "relationalities of our worldly engagements, both human and with other creatures,"[36] and here we include the things and objects of *Nature*.

Having discussed *spatiality* we now turn to *temporality* or what Wood refers to as "the plexity of time," which might be understood, in relation to the concepts of "complexity, implexity, and perplexity" as further related to the term "plexus."[37] Heidegger philosophizes a unique sense of *spatiality*, but he also, with his novel concept of *ecstatic-temporality*, provides a non-linear understanding of what we have termed existential-ontological "lived time," where temporality is instantiated in the inter-working and inseparability of the three temporal horizons (*ecstases*) of past, present, and future. Wood, in line with a reading that in some ways might be mapped onto Heidegger's phenomenology, claims that the everyday understanding of time as "linear," if we accept that time can be measured accurately through an understanding of "clock-time," obscures the deep sense of complexity and relationality that is at play when not only conceiving time as an ontological phenomenon, but more importantly, when *living it*. According to Wood, to live time is to at once experience "the continuous identities of things, the coordination of their processual and pulsing rhythms, and virtual and imaginative ways in which even in the instance we enter a connectedness that transcends the moment," and *temporality* sets us within experiences that are attuned by the "powers of interruption, interference, and breakdown."[38] Here, as related to our talk of "emplaced" dwelling, when *living* the "plexity" of time, the potential for *ek-static*, anticipatory moments of "standing out" with the potential for *transcendence* becomes an authentic possibility for our Being-in-the-world, which is linked inseparably with our more primordial *Being-in/with-Nature*. We now focus on four aspects of *temporality* that Wood identifies and develops: The invisibility of *lived time*, the celebration of finitude, the coordination of rhythms, and the interruptions and breakdowns experienced as *lived time*.

Lived time instantiates what is "invisible in the heart of the visible,"[39] for what is essential regarding *time* as *temporality* remains below the surface. The *lived time* of animals, mountains, oceans, and trees is invisible to us in everyday modes of inhabiting the world, i.e., the way things are temporally extended – as in *stretched out temporally*, persisting and growing – evades our glance, is occluded from us. Take for instance the temporal life of a tree: When we only focus on the tree as a material object that stands before us occupying space, envisioning it in terms of a "product" for consumption or obstacle to the pursuit of our mercantile desires, we miss the existential import of its *living time*. Indeed, it is even possible to experience the tree in terms of its component parts, its trunk, limbs, branches, or leaves, and here again, the existential understanding or experience of the tree as a living *temporal* entity remains hidden. In addition, we must note that when things in *Nature* change, grow, develop, and transform themselves, their essential ontological *presencing* defies the present moment and lives beyond accurate ontological representations. In direct relation to this issue, Wood discusses the temporal notion of the interruption, discontinuity, and breakdown, or the general unpredictability, of the manifold relationships comprising and unfolding as *Nature*, which in relation to *self-hood* and the *ek-static* essence of *transcendence*, embraces the fortuity of the potential arrival of what is unexpected and inexplicable, e.g., the miraculous, or a bestowal of good fortune. However, as Wood reminds us, there is also, bound up with what is unexpected, the potential of disaster and death, as in instances when unfortunate and tragic events are portended in the arrival of "the ghosts from another time that still haunts us."[40] Acceptance of this understanding should caution us against attempting to establish or conceive explanations of *Nature* that might represent a "premature holism" grounded in our "overenthusiastic drive to integration,"[41] or what we now recognize as the systematization of *Nature* as encountered in views embracing *scientism*. For as Wood rightly observes, eco-phenomenology seeks to authentically attempt to "*provide a model of the whole as something that will inevitably escape our model of it*."[42]

The so-called *celebration of finitude*, is the "concretization" of the present moment or the *intensification of the concrete* in our pre-theoretical experience of *Nature*, which is an intuition or attuned understanding that because *temporality* is given structure and boundaries by *finitude*, there is a fleeting and ephemeral essence to all of our encounters, which in one way, we might say, is an expression of the *fragile*, *unique*, and even the *precious* irreproducibility of our encounters – their essential *temporal singularity*; and so each such moment, we suggest, should be *lived-through* with a sense of gratitude and appreciation, as an "individual event" which "generates a kind of depth to every moment through which its very singularity is heightened."[43] Wood also claims that the experience of the *coordination of rhythms* of unfolding temporal phenomena, and the potential flourishing of *Nature*, is subservient and contingent upon the "advantages accruing from the eventual rhythmic coordination and differentiation,"[44] providing a temporary ordering and organization to *Nature*. Recall that this sense of organization in *Nature* bespeaks its integrity, its *inherent value*, e.g., fireflies, as Wood recognizes, light in a synchronic manner at evening's end, in step with the rhythm of the day, and the cicadas "carefully space (or time) the periodicities of their emergence from hibernation so as not to overlap and compete."[45] Wood's eco-phenomenology provides us with a vision of *Nature* in terms of a collective of manifold processes waxing and waning in concert, of entities and things interacting and intermingling, which occurs *temporally* as an "emplaced" or "spaced" and *emplaced* phenomenon. As Wood explains, because things "bear and embody rhythms, pulses of temporal development, they form part of a manifold and stratified field in which these rhythms interact, interpenetrate, interfere with one another,"[46] becoming either locally coordinated and united or differentiated and divergent.

Recalling how the transporting of trees from their "emplaced" dwelling to a foreign region disrupts and destroys the *originary* sense and experience of existential *spatiality*, we recognize that such instances of disruption also have a profound influence on the experience of *temporality* or the "plexity of time." To understand this, we do not have to turn to descriptions or accounts of the type of massive, sweeping, and concentrated deforestation occurring in tropical rainforests to get a sense of what we shall call the experience of lost "lived" or "*existential-ontological* time." Just consider the clearing or removal of strands or groves of trees that occurs on a regular basis when converting forestland to non-forest use, as in the all-too-commonplace practice of constructing urban and suburban strip-malls. This demonstrates the heedless disregard for the existential characteristics associated with an understanding of *temporality* and *Nature* as philosophized by Wood. For in this instance, the inherent *coordination of rhythms* that sustain the trees within the emplaced surroundings they share with other entities and dwellers of the forest – whose lives are ontologically dependent on each other – and that instantiate and enliven the *pulses of their temporal development* has been disrupted. Beyond this, both the *lived time* and *lived space* of the trees and the other entities that inhabit the forest, the overarching "emplacement" of the forest dwelling, is destroyed, and the shared forest that was once *lived* and experienced as the "manifold and stratified field" in which "rhythms interact"[47] is no more. Thus, we stress that it is not only the trees which suffer a dismal fate, for it is the unfortunate case that the entire local forest dwelling and its inhabitants are destroyed, for in the wake of *existential disruptions* to *Nature*, essential and *mutually sustaining relationships* are sacrificed. To those profiting from the construction and operation of strip-malls, the trees, the forest, show up (*come-to-presence*) as located within a particular three-dimensional space, they manifest, not as entities or things within their intimate "emplaced" context of dwelling, but rather as "obstacles" to the furthering of instrumental and mercantile aspirations, *present* in the "now" as objects, *present* in terms of serving instrumental ends; and in the case of deforestation, this signals their ultimate destruction along with the *time* and *space* of their dwelling. In contrast to this instrumental view of *Nature*, Bonnett provides for us a vivid description of the *living presence* of a beech tree in the woods, where *presence* is understood in terms of the "mutual interplay" occurring in the forest, for the tree occurs for us always in the

> interplay between the tree and its neighbours, the play of sunlight on its shimmering leaves, the rustle of birds flitting through its canopy, the aggregations of moss and lichen on its limbs, the enveloping odours of growth

> and decay. By participating in this interplay the tree both contributes to sustaining the place in which we come across it, and is sustained by it.[48]

Since we have explored *spatiality* and *temporality* as related to the human-and-*Nature* relationship, the analysis concludes by focusing on Bonnett's understanding of "human flourishing"; and to continue our elucidation of ontological themes, we relate *human flourishing* to *self-hood* and the existential understanding of *communality* or *lived-relationality*. The notion of "human flourishing," in its most recognized translation, comes from philosopher Martha Nussbaum and is derived from the Aristotelian understanding of *eudaimonia*, i.e., the ideal "good" toward which the ethical life should be directed.[49] However, according to Nolt in his critique of the "Aristotelian approach" in eco-ethics, it is reductive to base an ecological ethics on the notion of *human flourishing*, for it is related to, because it is centered on, the human being, and thus can be shown to harbor an explicit anthropocentricism. In succinct terms, addressing Aristotelian *Virtue Ethics* within the philosophizing of eco-ethics, we become more virtuous and excellent, with the potential to lead more meaningful lives, when directed toward "promoting or protecting nature's values."[50] At first blush this appears like an admirable and ethical path to pursue; however, as Nolt brings to our attention, in this instance the drive to protect and promote the flourishing of *Nature* is grounded in the following (questionable) logic: *If* we are better off by promoting, protecting, "or at least caring about the good of others," *then* we should demonstrate a concern for others. Extending this line of thought to include *Nature*, we then reason: *if* we are better off by promoting, protecting, and caring about *Nature*, *then* we should demonstrate a concern for *Nature*. The problem haunting this *hypothetical syllogism* is that the care for *Nature* primarily and essentially occurs because it benefits humans, and hence the obvious concern lies in the fact that although we seek to promote *Nature*'s flourishing, we do so in order to benefit ourselves, and this concern then holds the potential to inspire a form of "motivation" that is unconcerned with *Nature* in its own right, as it exists beyond us, as an intrinsically valuable primordial force. Ultimately, according to Nolt, although *virtue ethics* adopts "motive" as a crucial and legitimate criterion for determining "virtuous" behavior, the *directionality* of the motivation in this case is anthropocentrically inspired, centered, and grounded.

To reiterate what was established in the previous chapter, focused on ethics, the danger exists that if we adopt views that tend toward anthropocentrism, the best we can hope for is that humans *might* be moved to change their behavior toward *Nature* without necessarily requiring or searching out – let alone in terms of launching and embarking on a sustained philosophical or phenomenological inquiry – concerns beyond those that serve immediate *instrumental* purposes, e.g., as in *sustainable development*, where we encounter concerns related directly to long-term human survival and well-being. Nolt's foregoing critique of eco-ethical views grounded in *virtue ethics*, as is now obvious based on our eco-phenomenological exploration of Lingis, Bonnett, and Wood, does not take into consideration the deep, complex, and intimate connection that exists between the human and *Nature*. If we were focused exclusively on the human and its world, as is common with phenomenological investigations that do not include the environment or *Nature*, we would define "lived communality" or *relationality* as the "lived relation we maintain with others in the interpersonal space that we share with them"; indeed van Manen terms this existential theme or meaning-structure the "*lived other*."[51] In moments "when we meet the other we are able to develop a conversational relation which allows us to transcend *ourselves*."[52] However, as is our concern, the *other* includes the natural world or *Nature*, and so the form of *communality* that concerns us must include an analysis of *Nature*, for it is *only* within this unique and intimate relationship with *Nature* that an authentic sense of human *self-hood* and indeed flourishing, in terms of the potential for what we have termed *finite transcendence*, becomes a legitimate possibility for our lives. Let us keep in mind the indispensable essential message of van Manen's view regarding existential *themes*, specifically, that while they "can be differentiated," they cannot be separated or surpassed, for they "all form an intricate unity"[53] within our lives, in terms of giving structure and opening the potential for meaning in our *lived experience*, and it is for this reason that we demonstrated care in elucidating and stressing the all-important interrelatedness of the *themes*, for they all relate directly to grounding and shaping our understanding of *self-hood*.

As we have discussed throughout, for Lingis, the imperatives in things (*Nature*) are essential for the facilitation, formation, and sustained development of our sense of *self-hood*, and there is a complex relationship that he philosophizes that separates his view off sharply from both post-modern and classical empirical views of human *subject-hood*. As introduced, the imperative serves as the antecedent locus of human thought, perception, and subjectivity, and as Wheeler reinforces, our response to imperatives precedes our ability to "formulate principles with correct concepts and to reason rightly."[54] This of course, as shown in Chapter 4, is why the ethical imperative for Lingis cannot be grounded in a Kantian view of human reason, because it is the case that subjectivity originates in our subjecting ourselves to the force of the imperative, which as Wheeler articulates is our *undergoing* and *suffering* the imperative. However, his understanding transcends our typical use of "undergo," as indicating a "passive" experience or event to which we are subjected, in terms of our assuming the role of impotent or resigned "recipient." We must understand that for Lingis, "subjectivity is not only passive but [is also] active,"[55] and what this indicates, in an important way, is that the *anticipatory* attitude or mindset that opens us to the *demand* of the imperative, readying us for the reception of its *call*, first immerges as a "passive" inroad (potential) to the eventual "active" requirement for a decision, a commitment to choice and action (actuality) that is encountered in the *face* of the *call* and ethical *challenge* of the imperative. In the supreme ethical moment of decision and choice, autonomy manifests, for if subjectivity is to ever "become a fully sovereign force," it must originate in *subjection*, in terms of a "force in fortitude rather than a mere locus of empirical impressions or events, or mere construction of language."[56]

Wheeler also points out an important distinction between what is a *demand of power* and a *command of force*, as related to the imperatives in Lingis. We experience the *command of force* when in the presence of the undeniable and inescapable exigency of the ethical "force" of the imperatives; however, this is *not* to indicate that *subject-hood* can be reduced to or understood in terms of an externally "constructed subject," where we merely respond or react to an external *demand of power*, i.e., we are slavish and subservient, bowing in submission to a superior force, depending on that force *alone* for all meaning, a view wherein *subject-hood* and autonomy are either rendered impotent or radically put in question. To adhere to this latter model of *subject-hood* reverses the "relations of command and obedience," and this move destroys "the imperative's inherent hierarchy of command and subjection."[57] For as Wheeler observes, relations of power are normally reversible – as in the infamous philosophical example of *master-slave* relationship-dialectic in both Hegel and Nietzsche. However, the relationship of power we experience with the imperative is not: For Lingis, the ethical command of the imperative entails "an irreversibility that cannot be found in empirical or postmodern views of demands of the self as an organizing power as a bundle of sensations or a self-designation."[58]

Just as *self-hood* hinges for Lingis on our interaction with the *things of the world* and our various responses to the imperatives they emanate, the sense of *self-hood* in relation to *Nature* pervades the work of Wood and emerges in his idea of the "naturalization of consciousness," by which he means consciousness as it is always already immersed in the "natural world," and not as it is attuned by the *naturalistic* attitude (CF). This view is related to what we have termed "immersed-intentionality," drawn initially from Lingis's "corporeal intentionality" and Bonnett's notion of "emplaced transcendence," and indicates, as a running theme, that *Nature* is essential for and plays a crucial role in our humane and ethical development, where *self-hood* is in tune with *Nature*'s "cohesive integrity, which leads, eventually, to ways of managing," or responding and inhabiting "boundaries," where we encounter phenomena that inspire our thoughtful behaviors through our contemplating and appropriately responding to "distinct questions of meaning, identity, and value."[59] To engage *Nature*, and *engagement* here is the *communality* instantiated within the "dialogue" unfolding between the human-and-*Nature*, in terms of responding to its *needs*, as responsive to the sway, to the ebb and flow, of its pulsating rhythms, is to return to the very "conditions of [potential] human fulfillment and connectedness, but also to the sources of renewal, transformation, and resistance."[60] For Bonnett, what is crucial is not what Heidegger has expressed as the *Being-with-others* (*Being and Time*), but rather what might be termed our general and overarching "*Being-with-otherness*," which not only includes humans, but also non-humans, entities, and the things and phenomena of *Nature*. Human flourishing, an ideal we hold up for human *subject-hood*, is not "exclusively human" ("self-centered") because it is not purely a "human construction," for rather, in relation to our developing notion of *Nature*, it is a phenomenon necessitating the embrace of "non-human" aspects of our existence, and this, according to

Bonnett, is determinate of "human essence that, if expressed in the way we live [with *Nature*], constitutes our well-being, constitutes authentic dwelling."[61] This indicates, as is the main thrust of our analysis, that the implications for *self-hood* and human flourishing are grounded in adopting the right or ethically correct stance toward *Nature*, which instantiates what we have referenced as "authentic dwelling," and as we contend, this dwelling unfolds as a continuous and ever-renewed dialogue, grounded in and hence contingent on the various manners in which we respond, offering appropriate rejoinders, to the *address* and *call* of *Nature*; and so human flourishing, we then might say, is wholly contingent on the extent and degree to which those responses are appropriate in terms of being *ethical* and hence *authentic*. It is the anticipatory stance we assume in the face of *Nature* that is the essential key to authentic *self-hood*, which is a standing in the midst of a value-laden space, and according to Bonnett, as we have already encountered in Lingis's understanding of intentionality and *self-hood*, our "responsiveness to nature … is an aspect of our essence,"[62] and in reading this section in relation to Lingis's understanding of our encounter with the imperatives in the *things themselves*, we now understand what "immersed-intentionality" is and how it unfolds, along with the manner in which it instantiates our ineluctable relationship with *Nature*.

There is the potential, to reiterate, for our responsive and attuned stance to embody the awareness of what constitutes fitting and unfitting rejoinders to *Nature*, in terms of our ways of behaving that hinge upon an alert and careful *listening to and for the other*. Bonnett admits that finding the appropriate and fitting response is a "challenge of the highest order, as can be resisting the many powerful pressures to do otherwise,"[63] and we explore this notion below when tackling the crucial issue of the *difficulties haunting dialogue* as expressed by Lingis. Humanism, as we have already critically outlined, *does not listen*; rather it turns a deaf ear to the call and cry of *Nature*, or muffles and drowns out its call by chattering loudly, complaining vehemently, or pompously proclaiming with a deafening roar the superiority of the human race while proceeding to subjugate *Nature* for exclusive "human" purposes. For in the relationship we are describing, experience "is no longer exclusively orchestrated according to instrumental purposes and discursively constructed conceptual schemas."[64] To respond appropriately to *Nature*'s call or address demands what Bonnett terms a "thankful listening," which, as philosophized by Lingis, is *co-primordial* with and inseparable from the call of *Nature* and a form of attuned anticipation that recognizes *Nature*'s authority, and as we have seen, manifests in one way in terms of our admiration and respect for *Nature*, and "attempting to do justice by [*Nature*] and have justice done by it," as Bonnett recognizes, "is life's vital core."[65] To summarize, understanding *subject-hood* in and through our relation to *Nature* is to understand along with Bonnett, that our engagement with *Nature* "entails a fully bodied, multisensory participation," generative of our way of instantiating conscious-intentionality, which "involves feeling as much as cognition, and receptivity to intimations of fitting and unfitting response [to *Nature*'s call or address]."[66]

Readers are familiar with the etymological relationship between the Attic Greek "*oikos*" and the contemporary English word "ecology," but in relation to Bonnett's and Wood's eco-phenomenological ideas regarding *Nature* and its intimate "contexts" and rhythmically unfolding "interrelations," we can extend the sense of the term's meaning, for in addition to "house" or "abode," *a place of dwelling, oikos* also refers to the management of the household, as in the care and organization of the personal affairs of the "family" constituting the household.[67] Hence, we get the sense of *Being-at-home* with others with whom we share an intimate dwelling. In relation to our talk of *oikos*, Lingis astutely emphasizes the following: "A home is a zone in the depths of the elemental reality where reality is at rest," and when approaching existential contexts and boundaries, it is "established and maintained as a point of departure and point of return."[68] To have a home, according to Lingis, is to "actively care for the things appropriated," and here we learn that his understanding further extends our definition of *oikos* by insisting that the concept lives beyond the mere understanding of the physical abode of dwelling, the "house," and rather, in the mode of the infinitival, instantiates a dynamic process of caring for our space, our home or *oikos* within *Nature*. Lingis insists that we should tend and nurture this dwelling space, in a way that relates to our analysis of recognizing both ontological *spatiality* and *temporality*, by attentively caring "for the needs of the land, the plants and the trees, the herds and the flocks."[69] Along with the sense of *Being-at-home*, which for Heidegger is *Heimish*, indicating a sense of *coziness* and *hominess*, there is also the potential for experiencing the "*not-at-home*," and Heidegger gives what is perhaps his most moving and poetic portrayal of "homelessness" or the state of *Unheimlichkeit* (not-being-at-home) in his analysis of the tragic character of Antigone in *Hölderlin's Hymn der Ister*. Heidegger concludes that the essential, ontological place or dwelling of the human being (*Dasein*) is underway, is enacted, as an existence that instantiates *Being-at-home* in the plight or primordial existential condition of "not-being-at-home," i.e., the human's *most* original way of Being is that of "not-being-at-home," for it is ontologically condemned to continually seek to establish a "home" amidst a foreign landscape; much like the proverbial *Son of Man*, we never find a place to rest our heads while wandering the earth. Based on our analysis, the earth or *Nature* can never be made completely known to us, we can never assimilate it in knowledge nor fully master it through any technological means available to us. So, in this sense it cannot be stated that *Nature* is *one-with-humankind*, but this does not preclude us – nor should it deter us – from working to find and establish a home and dwelling within our natural surrounds, which occurs when we respond to those surroundings in and through what we have emphasized are *fitting responses* to the *call* or *address* of *Nature*.

Bonnett gives us an interesting take on the contemporary condition of *homelessness*, and here we return to what was presented in Chapter 2 regarding the CF, for Bonnett does not trace homelessness to the ontological and primordial condition of being *thrown into the world* as a human among what is and always remains as foreign, although what Bonnett philosophizes certainly might be said to represent a response to the primordial condition Heidegger identifies. Bonnett uniquely links contemporary "homelessness" to the *lived conditions* that humans have created, specifically, the conditions of life fostered and perpetuated by *secular humanism*. Because secular humanism "impacts upon the human need to find a home in the midst of the growing homelessness brought on by the humanistic impulse for mastery,"[70] where, and here recall Gray's vociferous critique of humanism and the dangerous illusion of unimpeded *human progress*, the destructive and suicidal drive to continue to dominate and subjugate the planet actually works "to deliver us ever more completely into the iron cage of technology."[71] To recognize *Nature*'s authority over us, and this is something we have already drawn from Lingis's phenomenology and his work on imperatives, need not "reduce humankind to the servitude of medieval or Renaissance view, but recognizes that humans have accumulated profound power, choices and responsibilities."[72] The question of human *self-hood* is grounded in *asking, listening*, and *responding* to the challenge of learning *how* it is that we can most appropriately and ethically accept our responsibility to make good choices in the face of *Nature*, which demands that we acknowledge the limitations of our knowledge, will, and power and admit that we should refrain from using the powers we do have in "ways essentially destructive to ourselves in nature."[73] Hopefully, there are some potential lessons addressing this eco-phenomenological challenge that can be drawn from the chapters of this book, inspiring in this and future generations, as Wood suggests, related to our discussion of "themes" as existential meaning-structures in our pursuit of the human-and-*Nature* relationship, the understanding that eco-phenomenology might contribute to an "enhanced attentiveness to the complexity of natural phenomena and the ease with which that is hidden from view by our ordinary experience."[74]

Listening to and for the "reticent" imperatives of Nature

Sonority and the essence of communication

Lingis, referencing Sartre, states that in the "look" (*le regard*) I "appear shameful or ridiculous or clumsy,"[75] for as I peer through the keyhole, approached from behind by the other, locked in her *gaze*, I feel ashamed and am transformed into an object, I am *objectified*, and a demand is placed on me to respond. As Lingis indicates, in relation to Sartre, my *subject-hood* (freedom) has been frozen or reified. Beyond Sartre, Lingis notes that when in the presence of the "face" of the other, as philosophized by Levinas, when *it* is *facing* me, (it) he/she is demanding a response. As we have seen, in a manner that transcends Levinas's ethics, Lingis contends that in addition to humans, the things of the world also wear and show their *faces*, and our encounter and experience with their faces, according to Lingis, "is already a linguistic situation. It's a question and demand."[76] In this section, we focus on the issue of how imperatives of the world or *Nature* manifest, not only, however, as organizing perceptual principles, as we have already explored, but also in their antecedent presence as phenomena of "language," carried in and passed along through *Nature*'s communication, which is instantiated in the "dialogue" that we *are* with *Nature*. In line with this pursuit, we develop the idea of "ontological aurality," an attuned *listening to and for the voices of things/others*. Engaging Lingis, we make the unique case that the relationship with *Nature* is grounded in, as opposed to "orality," or the ability to speak and sound language, our predisposition for "aurality," i.e., the ontological predisposition to hear and listen in anticipation for the call of others, world, and *Nature*. As Lingis asserts, we "speak" as a rejoinder to *Nature* through our responses and behavior. *Nature* speaks concerning its *needs* through its call or address, a medium of "communication" for the imperatives, which we learn represents the "inner noise" – or inner *ethical voice* – sheltered within the sounds of the waves crashing against the shoreline, the wind whistling through the tops of the trees, the birds and insects sounding their songs, and so on. At other times, *Nature* speaks through *silence*, e.g., the deafening silence inherent in the destruction following in the wake of an ecological disaster. To return to Levinas, in a way that relates to the phenomenology of Lingis, such moments of silence are vividly described as a "rumbling silence," which resembles the sound "one hears when one puts an empty shell close to the ear, as if the emptiness were full, as if the silence were a noise," and as Levinas continues, silence is "something one can also feel when one thinks that even if there were nothing, the fact that 'there is' is undeniable."[77] When describing and illustrating what this silence is like, it is interesting to note that Levinas references the encasement, *oikos* in the sense of a protective dwelling, or the shell of a once-living marine mollusk. What is crucial about this understanding of silence is that it is experienced as a resonation or reverberation within the inner recesses of our essential Being and not merely an *aural* phenomenon, an isolated act of "hearing" nothing. Silence, as related to embodied or immersed-intentionality, is a profoundly moving and pronounced *tactile* phenomenon. Lingis observes this and asserts that in silence, the "rustling, throbbing, whooshing, buzzing, ringings, and squeakings" that we experience, induce our muscles, our glands, to resound with the "ripples and rumbles of the never-ending movements of the atmosphere."[78]

Lingis contends that language arises out of the primordial *sonority* of *Nature*, from the "animal cry and murmur, outcries and stirrings, and remains rooted in this affective materiality," and this for Lingis only later, through a second-order manifestation of communication, serves as a "means of identification, commerce of ideal content and coded information."[79] It is in the very "sensuous density" of the things of the world, of *Nature*, that directives are sheltered, and "they emanate a savor, a resonance, an aura."[80] However, we must remember that Lingis is doing phenomenology, thus the claim is not that such an understanding is attempting to forward a rigorous, empirically defensible linguistic theory of the origins of human language, for even linguists do not know with certainty the empirical origin of language, but it does echo and intimate, in a far more *poietic* and thoughtful manner, the "hypothesis" regarding the origins of language found in Charles Darwin (*The Descent of Man and Selection in Relation to Sex*), who writes, "Language owes its origins to the imitation and modification, aided by signs and gestures, of various natural sounds, the voices of other animals, and man's own instinctive cries." Lingis discusses the many ways that we communicate, from writing to posturing or *body kinetics* – gestures, sighs, facial expressions, and so on – to speech given structure by grammatical and syntactic rules, all ways that employ signs and established conventions for communication. Lingis is focused uniquely on the sounds that are generated not only by humans and non-human animals, but also more broadly by the things of the world, inorganic and inanimate things, e.g., Lingis claims that when "we speak to another of a mountain vista, it is because that mountain landscape spoke to us," and indeed, *sonority* indicates that the world or *Nature* "teems with signals continually being broadcast from all configurations and all surfaces of things."[81] Examining the phenomenon of *sonority*, he recognizes that the focus of our attention, when posturing ourselves to hear something catching our interest, always occurs within a context of background noise, a context teeming with buzzing and humming static. Communication is the act of extracting a "message" from its background noise and then, as related to the call of the imperatives, further discerning the "inner noise" that instantiates the message being communicated. Communication occurs within a context or *space* where sounds are embedded, emplaced, and emerge from out of the places we inhabit and dwell; and much like the phenomenology of perception, when communicating we are actively separating a "figure" from its background, all the while acknowledging the necessity of that background, for as Lingis recognizes, "in the absence of the background there can be no figure either."[82] Imagine the leopards emplaced on the African savanna or mountain goats striding a steep rocky cliff with other animals, or a stalactite hovering from the roof of a cave among a host of other hanging structures, which includes the stalagmites rising from the cave's floor. Lingis argues against the attempt to construct a scientific and mathematical form of language or communication in order to "silence the rumble of the world,"[83] e.g., as suggested in the *isomorphic picture theory* of propositional language that the *Vienna Circle* extracted from Wittgenstein. Indeed, Lingis wonders whether or not such an objective, abstract, and scientific-mathematical language, if such a language were even possible, could ever really belong to anyone?

Lingis speaks of the looming and ever-present violence that resides at the heart of our attempts to communicate, for communication since classical times has been understood as *eristic* or in terms that are *agonistic*, e.g., there is a sense of violence that pervades the ancient Greek practice of forensic debate and contests of rhetoric. It is possible, of course, to understand communication in terms of mediating and quelling violence and mitigating competition, but it is more correct, according to Lingis, to understand communication as ineluctably bound up with *violence*, and not only in the sense of seeking to dominate and silence the voice of the other in the quest to establish the rightness, correctness, or superiority of our claim and argument. For there is also a sense of violence bound up with communication associated with the act of "preventing violence" in discourse, i.e., when "two individuals renounce violence and set out to communicate, they enter into a relation of non-communication and violence with outsiders,"[84] and Lingis refers to this view as "communication in the continuation of violence, but with other means."[85] More will be said about the "barbarians" storming the citadel of dialogue and the "outsiders," or *enemies of communication*, seeking to disrupt or silence communication. So, just as we have established existential *boundaries* in our so-called "dialogue" with *Nature* in our analysis of immersed-intentionality and the *ontological-existential themes*, it is also the case that all dialogue or communication has an "inside" and "outside," boundaries or *thresholds*, which should be neither traversed nor broached, but rather acknowledged and respected. In addition, as Lingis points out, communication also depends on finding the right "words" and "phrases" that we employ to communicate when passing along the experiences contained within the moments of our temporally unfolding lives. However, we note that communication is never reducible to the meanings attached to our words by certain codes that ensure communication, for it is also the delivery of our words, the cadence of our speech, that also carry and convey meaning, for communication instantiates the visceral and embodied "rhythm [and] tone, the periodicity, the stammerings, and the silences"[86] that are essential to communication. Importantly, we acknowledge that in addition to those who would silence communication, the *enemies of communication*, there are also allies, e.g., philosophers, eco-phenomenologists, and other environmentally concerned thinkers, whom Lingis refers to as "outsiders who have an interest in promoting the communication between us,"[87] or as in our case, between humans and *Nature*. Indeed, by inviting the reader into this book with an anticipatory attitude or mindset, as an *ally* in furthering communication, we hope to promote and inspire an understanding of authentic dialogue with the natural world and welcome the reader as an *ally* in the quest to re-conceive, renew, and expand the authentic and sustained ethical communication between the human and *Nature*.

We note, for readers encountering Lingis's phenomenology for the first time, that when writing of communication and the murmur and noise of the world, we are merging two sets of ideas and lines of argumentation from *The Community of Those Who Have Nothing in Common* and the later phenomenological text *Imperatives* as related to our exploration of ethical imperatives of the world as explored in Chapter 4. This is crucial because we are making the case that the "voicing" of things, or the *call* or *address* of *Nature*, instantiates our "immerged" engagement with the imperatives in things and the subsequent prima facie *duties* that emerge and to which we are bound by a primordial ethical duty. To reiterate, the imperative's directive is sheltered within and inseparable from the sensuous depths of the things of the world, as "attractions … they draw our perceptual movements to themselves and hold them," they communicate their messages and place demands on us as they mark out the "terms" of our relationships – through a voice that strikes *at* and *to* the heart of our Being – and these *terms*, which establish the *originary* parameters of our ethical engagement, according to Lingis, "rest in themselves, termini,"[88] as they are intrinsically valuable and binding. We now understand that humans face each other and their *faces* speak, and that so too does the world of *Nature* speak, emitting in and through its voice and the countenance of many *faces* "signals that can be received, recognized, and reiterated," when understood and taken up with an authenticity that instantiates the responsibility of our duty to honor the ethical commands these voices communicate, commands which sound and echo within a context, within a *space* comprising the "humming, buzzing, murmuring, crackling, and roaring world,"[89] the *lived-world* in its primordial and brute nakedness. When describing the *call* of the world creatively and poetically in terms of *Nature*'s "music," Lingis provides a moving account of a native's dwelling in the rain forest immersed in the phenomenon of *sonority*, where

> the Indian hears the animal, vegetable, mineral, and demonic realms. One had to listen to it there, in the nights of the Lacandone rain forest, to understand that this "music" is not an aesthetic production, that is, a creation of human subjectivities attempting to communicate immanent states like moods, feelings, values, or messages to other human subjectivities. It is a prolongation of the forest murmurs, the whispering sands, and the hum of the heavenly bodies.[90]

Considering the sense of responsibility that accompanies our reception and appropriation of the imperatives of *Nature*, authentic communication, for Lingis, is irreducible to a mere medium for passing along information. For there is an intimate essence to both communication and discourse, and one "enters into conversation to become an other for the other," as now might be related to our understanding of *self-hood* and the potential for *finite transcendence*, which can be *understood* in terms of the intimate and ineluctable connection between our life and the other's life and the dedication *to authentic communication* that "accompanies the harkening to the message,"[91] with the "singularity and our indefinite discernability" being instantiated in and inseparable from "our outcries and our murmurs, our laughter, and our tears: the noise of life."[92] We have talked of appropriate rejoinders to the call of *Nature*, when the world of *Nature* speaks to us and we take that call and address up into our Being, which for Lingis means appropriating "the [sonic] vibrancy of the land, the oceans, and the skies," which is "condensed, and unfurled in the hollows of one's body, then released,"[93] and when attuned, we hear the echo of our rejoinder returning to us across the land, over the ocean, and from out of the brilliant skies. In the *returning echo* of the other, of the things of *Nature*, if we have adopted an appropriate stance, offered a fitting response, we learn through *Nature*'s sonic approbation that we have done right by it, that our responses were appropriate to the demands placed on us. As already introduced and as we further elucidate below, what is vocalized and communicated through the call of *Nature* is its visceral, recognizable, and "discernible vulnerability," which intrudes on us, penetrating our world, disrupting our complacent and disengaged ways of inhabiting the space of *Nature*, an intrusion that holds the power – if we allow it to "speak" to us, to *attune* us – to potentially shake us from our dogmatic slumbers as it reveals to us its *needs*, its *fragility*, and ultimately, as Lingis contends, its potential for death or *mortality* – as yet another *ontological-existential* structure – to which all things are inevitably susceptible, and which is what Lingis refers to as the *summons of death*. For in communication, in our dialogue with *Nature*, in the strengths and weaknesses of all things human and non-human, there is a sense of *mortality*, and we perceive this as the possibility of *Being-no-more*, and ultimately, we are moved, indeed overwhelmed, as we feel and experience, in the grip of the imperatives of *Nature*, "the weight of death in [all] things"[94] pressing down its ontological weight.

In this section we reference "reticent" imperatives, and yet thus far we have primarily discussed, save for our excursion into silence, the *sonority*, the vocalization, and the undeniable "noise" emanating from things. What does it mean to speak of "reticent imperatives," for most certainly our description of the forests, creeks, and animals highlights their ability to verbalize their internal resonations, their internal voices or "noises"? If we examine the term "reticent," we see that it can evoke the idea of a thing's "propensity to be silent, a disinclination to speak readily"; or it can indicate "having a restrained, quiet, or understated quality."[95] We must be clear that what is "reticent," however, is *not* the address of *Nature*'s call, its sounding, but rather, in line with a phenomenological reading, it is the imperatives themselves when communicated, when passed along, that are without a distinctly audible voice and so are *reticent*, for their message is hidden below the surface "noise," as it were, as it is sheltered within what is sonorously vocalized, residing within and behind the murmurs of the forest, the bestial wails of the animals, and the whimpers and last audible gasps of the dying forests and shorelines. This is what Lingis has in mind when contending that the "medium" carries "signals" that are continually being sent from the surfaces of things. These "signals" represent the thing's essential "internal noise," i.e., the medium carrying "messages" of the reticent imperatives that remain unintelligible to, as related to the focus of Chapter 2, those who are disingenuously attuned within the dominant influence of the *naturalistic* attitude. For if our focus remains continually fixed on the surface phenomena of audible sounds and beyond, if we wrongly believe that things simply emit nothing other than meaningless and distracting "noise," then we take a place among those who refuse to really listen, who do not *want* or *care* to listen to the internal noise of things, to the reticent message of the imperatives communicated through the internal resonating tone that is sheltered within the immediate, discernable, and recognizable "surface noise," e.g., "the rasping or smoldering breath, the hyperventilating or somnolent lungs," all of which harbor and communicate, to those who are attuned listeners, "the rumblings and internal echoes."[96] A rational and mathematic form of abstract communication, as we have stated, seeks to transcend the internal noise of communication, and in addition, "it has also not understood – has not *wanted* to understand" the background or the internal noise "in the midst of which we speak."[97] Both the "internal noise" contained in the call and the "background noise" are crucial to communicating the imperatives, for like our understanding of the existential context of immersed intentionality with its boundaries of *spatiality*, communication – as inseparable from immersed intentionality – occurs "out of and in the midst of the reverberation of ambient materiality," and the utterances formed by us in response to the things of the world are sent out and return to the context of background noise, a context that, as we have seen, is given structure by *spatiality* with its boundaries and *thresholds*, formed by the movement and sway of the rhythmic unfolding of *temporality*, where the reconceived understanding of *communality* unfolds through the dialogue between the human-and-*Nature*. Immersed intentionality in relation to Lingis's notion of communication makes it possible to understand that we are linked in the ever-renewed process of "struggling, together" to confront and challenge the "equivocal voice of the outsider,"[98] which seeks to eliminate all noise or to discount noise as something that is foreign, disruptive, and antithetic to authentic communication, and which, as stated, represents the *enemies of communication*.

A most obvious example of *Nature* communicating imperatives is something that might appear inconsequential, unless you are a caring pet owner, and that is the contented purring of a cat as it is gently being stroked, or if a bird owner, you are familiar with the soft cooing of contentment, reminiscent of a low purling murmur, that is common to many parrots. Through these types of sounds the animals communicate to us that we are responding to them appropriately as attuned to their message carried within and communicated through the imperative that frames the ethically fitting response in this situation, responding to the *need* to be cared for, for us to honor and enact our duty to provide what is required for a safe, healthful, and flourishing existence. In line with our analysis, it is not merely the audible sonic patterns and vocal cadence produced in the midst of their contented state, emanating from the animal, that directly communicate the imperatives of either *beneficence* or *fidelity*. It is also not exclusively through posturing and bodily adjustments made in such intimate moments that imperatives are communicated, for this phenomenon is also irreducible to the physical, although gestures, as a primal expression of embodiment, for Lingis, do contribute to the communication of inner Being and essence. Rather, the contentment we described above manifests and emerges as an "inner voice," contained within and transported by the animals' immediate and audible sounding. This "inner voice" authentically "speaks" to us regarding the stance that we *should* take, are *obliged* to take, when attuned to the presence and secretive transformative power held within the countenance of the animals' *faces* and released through their *call* and *address*, which communicates the reticent ethical imperatives for shaping our responses. This occurs, as stated, within a context or *space* given shape and boundaries by the interplay of sound and silence, where background noise is contained by the thresholds established by *existential spatiality*, and it is within these boundaries that we are able to listen and discern the reticent messages carried and contained within these specific soundings as they stand out, in bold relief, from the droning and obscure noises that tend to overwhelm us or pass unnoticed in our everyday ways of comporting ourselves in the world. So, it is possible to conclude from Lingis's phenomenology that authentic communication lies and occurs within the "zone" from out of which the imperatives speak to us, an ontological space or place where the communication of the message of the imperative silently resides, and must be wrested from concealment; communication of the ethical imperative resides in the ontological space of what we have already referenced as the notion of being-at-home in the "home zone," as a place where *reality rests*. Based on our analysis of Lingis, it is possible to understand the phenomenon of the communication of the imperatives of *Nature* as establishing our most primordial mode of dwelling in Nature, setting us *in-between* transcendence and finitude, *in-between* a life closed off to communication and a life where the potential for authentic communication becomes a legitimate possibility, if we are *attuned* to listening to others and *Nature* in ways that live beyond anthropocentric views of inhabiting and responding to *Nature* that foreclose such reconceived possibilities.

In these moments as described, we encounter imperatives opening us to our potential prima facie *duties* to *Nature*, and as we know a plurality of duties can exist within a single situation, for as *potential* our duties are neither objective nor indelible in nature. For example, as briefly stated above, consider our duty to *beneficence*, to treat the creatures of the natural world with kindness, charity, and respect, or as in the case of our pets, with love, while fully recognizing the animals' capacity for potential pain and suffering. We might also experience our duty to *gratitude*, to demonstrate and hence communicate our thankfulness and deep appreciation for *Nature*, for our animal friends that unselfishly give the *gift* of devotion and companionship. However, we can also imagine imperatives that signal the unethical treatment of animals, both domestic and wild, treatment that heedlessly ignores the imperatives sheltered within and communicated through their cries, howls, and wails, expressive of their extremely fragile natures, their potential for harm and injury, the potential for the degradation of their lives and ultimate deaths. Thus, we are also gripped by the cries, groans, and bestial wails of animals in peril, animals that are poached and destroyed for their hides or ivory tusks; animals that are heinously and senselessly tortured and abused; animals suffering the bleak and terminal fate of the factory farm system. In the presence of these types of calls and addresses we are faced with imperatives opening us to the prima facie *duty* of *justice and equity* toward *Nature* as well as our duty to *non-malevolence*, in that we should always resist doing evil and attempt to ward off evil wherever possible, avoiding malicious behavior and actions that do unnecessary harm to *Nature* in all of its sublime grandeur, in all of its various forms and manifestations. As introduced earlier, there also calls transmitting distress signals that erupt in a deafening and "rumbling" silence, such as the disturbing and ominous silence permeating the aftermath of an unnecessary and preventable environmental disaster that has engulfed the barren natural landscape once teeming with life.

Let us return once more to a familiar event and theme, to examine the many and varied "reticent" imperatives of *Nature* emanating from the Alaskan coastline following the aftermath of the EVD. However, prior to approaching the experience of the EVD and the way in which the call of *Nature* manifests, let us reexamine the eco-phenomenological questions, which were identified as the type of "original" questions that classical phenomenology asks, as we encounter them in their tacit expression in Lingis's phenomenology. Imagine what might be a pre-reflective or *originary* experience of a pristine coastline, untouched, and as of yet, undamaged by the human's presence and intervention; and so we consider the following questions in order to potentially return to and experience what we might call, borrowing from Harman, the "coastline effect":

> How does the coastline appear to us in its brute and naked originality when we have suspended the conceptual lens of naturalism? How do we experience the coastline in what might be understood in terms of a primordial experience? How do we interact with it when simply standing and admiring in its presence? Is there an experience of involvement that might exist beyond the aesthetic effect of the coastline along with all of its various life forms that might move us in the face of its beauty and majesty? Or, might an aesthetic view capture something that remains elusive and is absent from other ways of viewing? How does the coastline interact with and bring together its surroundings? Is the coastline's essence reducible to a thoroughfare for oil tankers transporting crude forms of energy, or is the coastline's Being, in its most original essence, something beyond this? What might it want us to know about it? Does it appear as value-neutral or value-laden? Does it exist to serve our needs, wants, and goals, or for some other, perhaps unknown, and hence unnamed purpose? Is it dependent upon us or are we dependent on it? Or, are we, in some important manner, in some degree or form, co-dependent?

As eco-phenomenology claims, asking, considering, and later reflecting on such "original" questions holds the potential to open us to, by putting us in touch with, a transformative way of *understanding* – or intuitive insight – that is revealed through our pre-reflective immersion in, which highlights our involvements with, *Nature*. Such moments we have said represent, embody, and instantiate our *attunement* in ways that set us outside and transport us *ek-statically* to a realm beyond the overwhelming influence of the *metaphysics of presence*. As discussed throughout, this is the unique experience of the power and inexplicable "effect" of *Nature*, in terms of its *commanding reality*, in the *face* of its grandeur and integrity. In such attuned moments, *Nature* is released from its servitude to *naturalistic* bonds, it is *de-fetishized* and rises up and above us as *idol*, and its "inner voice" speaks as never before with a crystalline clarity, in such a way that *Nature*, to once again invoke Harman, assumes its essential form as an "imperative, radiating over us like a black sun, holding us in its orbit, demanding our attention, insisting that we reorganize our lives along its shifting axes."[99] As we have stated, *Nature*'s voice, call, or address remains inaudible for the most part, because it is obscured within our everyday involvements with the world, and requires an eco-phenomenological method or approach to awaken us – *attune* us

– to the *originary* power of this voice, by clearing the ground for its release and emergence, for the *presencing* of its "inner voice" with its ontological sonic reverberation.

Being present (*Being-there*) on the shores of Prince William Sound in a pre-reflective mode, experiencing the raw integrity of *Nature* as we are immersed in the *space*, entrenched in an *axiological* mode of immersed-intentionality, we are temporarily free of the *naturalistic* attitude. What do we experience, and more importantly, what are we open to *hearing*, what do we *encounter* if we truly *listen* with our entire Being? If we *listen*, might not we hear and feel the reverberation of the unimpeded flow of the powerful ocean water crashing loudly against the shores of sand and rock that teem with life? Might we not also hear from overhead the piercing screeches and screams of the waterfowl that hover in the wind hunting for food in the clean, pristine waters? Other birds are scavenging on the beach, with only a thought for non-human predators – can we hear them signaling to the familiar and warning others away in their attempts to challenge them for their spoils? Being there, within this space amid all of the "noise," we discern the exigent messages sent our way through the imperatives, which urge us to consider seriously that the *space* and place of this pristine and undisturbed seascape and coastal land could always be otherwise, and so already the "internal noise" of the imperatives of Prince William Sound rise from out of the "background noise," carrying messages, and we are attuned to and moved by the essential condition that all of this beauty, all that is intrinsically valuable, is temporary, fragile, and exceedingly vulnerable, grounded, as Lingis contends, in the *summons of death*. Imagine now that we hear the Exxon Valdez tanker slicing through the waters of the coastal area, tearing through the waves with its engines roaring and echoing like a sonic boom across the horizon; and when this *echoed* sound returns from out of the distance, what message does it carry? There is already undeniably something *foreign* here in the midst of the *familiar*, making its unwelcome and unwanted presence known, intruding into the *ontological space* and disrupting the boundaries and thresholds that *Nature* has established, bringing with it through its "sounding" and *address*, as was the case on that fateful day in 1989, the ontological *summons of death*, which served as the ominous harbinger of the premature arrival of death in the *all too real* eradication of mass species of mammals, fish, and birds, i.e., the disruption and destruction of the ontological *spatiality*, *temporality*, and *communality* of this coastal dwelling place.

Now, as opposed to the sound of the waves crashing against the shore, bringing sustenance to the land, gurgling and frothing up before returning to their origin, the waves sound a death knell, a muted drone that has been choked out by gallons upon gallons of crude oil gathered on the surface forming a thick, viscous blanket, and as opposed to freely flowing and washing over the shoreline, they now appear to slowly creep and slither onto the shore, covering the sand and rocks and leaving a heavy black reminder that *Nature* has been defiled; and again, we experience the *intrusion* of the *foreign* in the *familiar* with devastating results. Then, let us further consider the fate of the sea birds that in many cases will call out one last time before succumbing to the "rumbling silence" of death. Birds, trapped at sea by the heavy oil, weighed down and submerged, violently flap their wings, slapping the water's surface, creating the desperate sound of their imminent demise as they gasp loudly, struggling mightily for their final breath, and in their call, in their soundings, their terror and desperation is given "voice" before the black water envelops them and drags them down below the surface of the thick and viscous ocean water. This is because "birds lose their natural buoyancy from air pockets created by proper feather alignment, and they sink and drown in the polluted water."[100] Others die a slow and syncopated death over time from the toxic oil they have inadvertently ingested, for "instinctively, the bird tries to get the oil off its feathers by preening, which results in ingesting the oil causing severe damage to its internal organs."[101] In all of this, looking back from our current place in history, we are contemplating the call and address of *Nature* in moments that appear to come too late, but there is hope. For although it is impossible to travel back in time, it is possible to learn for the sake of the present and future by reliving the effects of the disaster, as difficult as this can be, by consulting history, by attuning ourselves to what is communicated through accounts of the disaster that include the gruesome and disturbing photos of the carnage left in the wake of the Exxon Valdez tanker that ran aground, all the while understanding that the residual and deleterious effects of that man-made disaster continue into the present moment. In this instance, as might be obvious, *Nature* might be said to urge us to subject ourselves to the imperative and the prima facie duty of *reparations* that is communicated, which is the duty we have to *Nature* and others to attempt to make amends and offer recompense for wrongs that have been done, whether this harm has occurred with purposeful malice or is inadvertently inflicted.

Prior to addressing the *enemies of communication*, let us pause to consider a crucial running theme in Lingis's phenomenology regarding the *transformative* experience of receiving the imperatives. As shown, the imperatives strike us immediately as we *feel* or *intuit* the *neediness* and underlying *destitution* of the other, and this for Lingis, relates to the pervasive influence of *finitude*, or what he ultimately identifies as the inevitable *summons of death*. It is the case that we cannot elude or outrun death, and in the "measure that we cannot retreat behind the pain, our vision is invaded by it and adheres to it," in all of our sufferings we struggle to "bear the pain" through "endurance and patience."[102] Just as the prima facie duty to *non-malevolence* might be said to serve as a unifying existential theme among the seven potential duties proposed, so too might the *summons of death* in Lingis's phenomenology be understood as the foundational ontological *theme* underlying and grounding the imperatives, substantiating their exigency and binding ethical force. As in our description of the EVD, we experience the *summons of death* in a moment of what Lingis calls "attraction," in that the imperatives demand a response to all forms of defilement, corruption, and decomposition. This for Lingis is simultaneously a call to a "destination," serving as the harbinger for our necessary and fateful life-task, and this vocation is the quest for authentic *self-hood* in relation to the imperatives.

However, as opposed to inducing a sense of *dread* or plunging us unto the squalid, fatalistic state of *passive nihilism* that Nietzsche argued so vehemently against, the looming potential of death "quickens our forces and summons them upon the pleasures the environment offers us," inspiring us to attempt to facilitate the prolongation of such pleasurable moments, in ways that require, as is now familiar, ethical responses to the "tasks it urges upon us, and the gifts of our forces for which it presents occasions."[103] An inevitable ontological certainty, death permanently cuts all ties to *Nature* and others, it spells the end of our immersed-intentionality, it terminates *spatial* and *temporal communality*, transporting us, as it were, "into the outer regions where the order of things and of the world enters into decomposition."[104] Perhaps, as Lingis speculates, death might be the ultimate encounter with the *sacred*; nevertheless, we enter into that realm *only* by losing our world and ourselves. Responding to the imperatives of *Nature* that carry with them the *summons of death* situates us, if we are attuned, alongside the other's vulnerability and potential for her/its death, and in such moments, if we are open to *listening*, *hearing* and *responding*, Lingis asserts that the potential for true "courage" becomes an authentic possibility: "Courage is the strength of heart that hears the advance of death as a summons to engulfment in the elements, to the horror of decomposition of a corpse, to extinction," providing the "steadfast strength to hold oneself together"[105] in the face of the other's and our own potential for non-existence, for death. Courage, it is possible to state with confidence, is then required for eco-phenomenology's quest to transcend the *metaphysics of presence* and reconceive and re-establish our lost and obscured relationship to *Nature* in all of its primordial power and force.

The dangerous "enemies of communication" with Nature

We now explore Lingis's dire warning regarding the enemies of discourse or communication as related to the elements expressive of the essential traits of language and communication, phenomena manifesting within the context or *space* of the "embodied" and "immersed" relationship we share with *Nature*. Thus far, we have provided analyses of the following three traits of communication, which are tacitly present to Lingis's phenomenology, but which have been made explicit for the analysis: (1) We explored the basic attunement to *aurality* as a defining ontological component of our Being-in-the-world, which we have seen, includes *communality* with *Nature*; (2) We analyzed the anticipatory openness required when releasing ourselves over to the potential of being changed and transformed ethically by the other's address; and (3) We looked at the crucial issue of the exigent need to develop an attuned sense of careful "listening," as in demonstrating a solicitous concern for the other's well-being in light of her/its *neediness* and *fragility* as vocalized through messages transmitted through the imperatives. However, we have yet to elucidate what might be labeled the *fourth trait of communication*, namely, that dialogue always has the propensity and possibility of being interrupted, dissembled, and terminated, or silenced, shut down by "outsiders" or by *barbarians storming the citadel of communication*. This issue was introduced initially when speaking about the air of violence inherent in communication. So, in what serves as a concluding section to this chapter, we address the necessity of *vigilance*, of always being on guard for the *enemies of communication* in all of the pernicious and insidious forms they assume.

For example, in instances where we attempt to make ourselves superior to the other, or to subjugate the other to our position, our will, Lingis claims we assume the role of an *enemy of communication*. This indicates that we are seeking only to establish the superiority and correctness of our position, this despite the strength of the argumentation, with the goal of silencing or closing down authentic discourse by drowning out or silencing the other's voice with "noise," and in the process, maligning her *subject-hood*. Here, it is necessary to remember what Lingis has said regarding the necessity to subjugate ourselves before and to the other, to *Nature*, as we assume an attuned stance before the imperatives, whereas the enemy of discourse seeks to reverse this ethical approach, or forestall it, e.g., we have already encountered this disingenuous trend in *secular humanism* in relation to *Nature*. Lingis also refers to enemies of communication as "barbarians," this because, we might say, just as the ancient Greeks recognized, *barbarians* are outsiders because they could not think or speak "Greek," and thus, as "outsiders," they remained outside of what we have termed the existential *space* of engagement with the *other*, and these are ways that express the contention inherent within communication, not in terms of the *back-and-forth* of discourse, e.g., as would be consistent with the *question-rejoinder-refutation* of the Socratic *elenchus*-dialectic, but rather the competitive contention that risks destroying the discourse and along with it the authentic potential for communication. We must also be aware of those who insidiously and subversively attempt to shield their discourse from outside criticism, e.g., when those with hidden agendas and ulterior motives vehemently defend the "established order, some established sense of values, or some veiled interests." These enemies of communication, in the effort to protect and guard their positions, are always behind the door, peering into the keyhole, listening to our conversations; they are "the enemy, a big brother, the little Hitlers."[106] With their own hidden agendas and nefarious ends, they destroy authentic communication by "filling the time and space with irrelevant and conflicting messages, with noise,"[107] with so-called "alternate narratives" or "alternative facts," or with outright and blatantly dangerous lies. For example, we currently experience as a common and rigorously documented practice of the Trump Administration in the US, how this extends to the EPA in its current configuration, for under GOP (Republican Party) rule it is composed primarily of climate-deniers, big-business supporters, and those eschewing and ignoring outright the conclusions and findings of the scientific community as well as the warning of global destruction communicated through various environmental philosophies. Such politicians and closed-minded administrations are true enemies of communication, for despite the scientific findings that warn of the dangers of charging ahead in the heedless overturning of policies and regulations that serve the end of conservation and the ratification of policies, they demonstrate an ignorant, undeniable, and disgraceful lack of concern for the natural world, and we have already discussed the intricacies and specifics of these issues in Chapter 2.

Lingis recognizes that the disingenuous view of technology, emerging from the attunement of the *metaphysics of presence*, intrudes into the boundaries of Human-and-*Nature* relationship. In relation to "noise," this view of technology might be understood as manifesting on two fronts: (1) The loud and deafening "noise" produced by the physical machinery used in the many and varied technological processes employed to bend *Nature* to the human's will – a human-produced, and hence artificial, "noise" that drowns out the call of *Nature*, impeding the authentic unfolding of dialogue, and (2) Those who endorse the disingenuous view of technology and cling to it – the "prosopopoeia of noise" – seeking to drown out the "noise" of legitimate communication – when it is directed toward opposing their dangerous positions – that represents positions such as ours that are antithetic to dogmatic, close-minded views that embrace the erroneous view, with a ruthless tenacity, that *Nature* is subservient to technology. Intrinsic to this latter concern is the move to ignore the call of *Nature* communicating the prima facie duty to *fidelity*, or the faithfulness to our obligations toward *Nature*, which is ultimately the responsibility we must bear to adhere to the imperatives of *Nature*, i.e., the *fidelity* to the inherent value of *Nature* and those who recognize it and seek to uphold it and argue vigorously for it. To this issue, which has been documented and confirmed by major news outlets across the globe, President Trump and his administration continue to concoct and spew lies in order to create false-narratives to combat and silence the *good science* that documents the dangers of proceeding headlong into rolling back and rolling out policies that are clearly destructive to *Nature*. For example, Trump has vowed to revive the coal industry in the US and, as stated earlier, is in the process of rolling back EPA regulations that would restrict certain types of mining because of their proven harmful and destructive effects on the environment. This includes reviving Mountaintop Mining, which is nothing other than a euphemism for the process of employing explosives to blast away top layers of mountains to extract coal, the underground sources of which have been exhausted, leading ultimately to the death of the mountains, local wildlife, and large swaths of forest land. Of course, the administration is unconcerned with the scientific data verifying the myriad of dangers associated with this form of mining, and so seeks to ignore, and beyond, "silence" such information.

Here, in addition to the *prosopopoeia of noise*, we must consider the *technological noise* of the sonic blasts occurring with a staccato-like regularity when mining mountaintops, disturbing both human and non-human dwellers in the area, drowning out the sounds of *Nature*'s communication. It is also possible to experience the "rumbling silence" as discussed, e.g., when the work has shut down for the day, and most prominently, when the coal source has been exhausted, the mountain site has been decimated and leveled, and the workers have moved to another site to continue their deadly method of excavation. There is the silence of the absent trees that resulted from the deforestation taking place prior to cutting into the mountains with explosives resounds and reverberates; the silence of the wildlife no longer present to the region, pushed out by the incessant noise and the danger posed by flying debris resulting from the explosions; and the haunting din of the *silence of the dead* – the wildlife poisoned in their once secure habitats by the chemicals and toxins from the excavation process, which silence streams, rivers, and grazing areas and, as related to humans, seep into the groundwater with the effect of inevitably harming and silencing human beings dependent on water supplies for survival, something which, as of this writing, is occurring in the local communities of Appalachia where this form of mining takes place. The loud and deafening explosions created by the technological processes of mining drowns out the "noise" of *Nature* in such an extreme and radical manner that it destroys any possibility of authentic communication; as Lingis laments, "by eliminating the tactual background signs, one no longer senses the boundaries between outside and inside, past and present, perception and images, one soon hallucinates."[108] This technological "noise" disrupts the very essence of the dialogue or *communality* between human-and-*Nature* that we developed above, in terms specifically related to *spatiality* and *temporality*, obliterating the necessary ontological boundaries and *thresholds* that facilitate and sustain authentic communication between the human-and-*Nature*.

As is obvious thus far with what we have stated regarding the majority of the current GOP (Republican Party) rule in the US, they represent, in a most vile and reprehensible manner – and not just as related to the environment or *Nature* – sworn enemies of dialogue and authentic communication. To conclude with one final "real-world" example of how communication is destroyed by the intervention of anthropocentric political policies and destructive technological practices, grounded in the advancement of the Nation's GNP above all other concerns, let us consider the process of seismic blasting on the ocean's floor in order to locate new oil and gas reserves off the Atlantic coast of the US. This once-banned practice, now approved, will undeniably do harm to thousands of sea dwellers, including whales and dolphins, creatures, we note, that communicate through sound waves, through sonar echolocation in order to find food, avoid danger, and navigate their habitat in advantageous ways. However, the danger is not limited to these two species: Also in danger are "sea turtles, fish and zooplankton, a critical foundational plank of the ocean web."[109] As Beth Wellington informs us, the process consists in giant seismic air guns firing continuous blasts, sometimes in intervals of 12 seconds, for prolonged duration, at times for months on end, in order to locate these fuel deposits. The deafening noise, and we must keep in mind that in the depths of the ocean, sound, which is monumentally amplified, can travel up to 2,500 miles. Here again, we are confronting the "noise" drowning out the call of *Nature*, and in seismic blasting noise levels can reach "260 decibels," which is equal to being located in "the epicenter of a grenade blast and would easily cause the rupture of the human eardrum."[110] All of this preparatory work is the harbinger for dirty and dangerous offshore drilling, the hazards of which we have already spoken of at length, and thus we encounter a prime example of how data regarding destruction to animals, their habitat, and the surrounding shoreline is ignored, i.e., the "noise" of empathetic reason, indeed the "voice" of empirical science, is stifled and subjugated to an unethical counter-narrative that ignores the ever-present danger to the destruction of *Nature* to which such technological practices inevitably lead. All of this we relate to the important and invaluable work of Bonnett, Wood, and Lingis, for the destruction to the dialogue that is *Nature*'s unfolding in all of these foregoing descriptions of the human's technological intrusion into the sacred space of *Nature* represents in ontological terms the invasion of the *foreign* into the *familiar*, where *boundaries* are traversed and obliterated, where the dwelling *space* and the *time* of its rhythmic unfolding are heedlessly destroyed, trampled underfoot in the incessant, and according to Gray, what will prove to be the inevitably fatal, march of human progress.

For Lingis, we enter into conversation with the other, the human, the non-human, *Nature*, only when we lay down our weapons and throw open our defenses and "throw open the gates of [our] own positions, to expose [ourselves] to the other."[111] When standing before the imperatives of *Nature*, subjecting ourselves to the *needs* and *demands* that they express, in such anticipatory moments of attunement, we release ourselves over to "surprises, contestation, and inculpation,"[112] and in this we can learn from *Nature*: (1) About the fitting and ethical responses that are called to make in order to live in a renewed fashion within the natural world and amongst all the entities that require our attention, respect, and assistance, and (2) About ourselves – by attentively listening and formulating a thoughtful rejoinder to *Nature*'s many and varied "reticent" imperatives, we are moved to respond with *contestation* and *inculpation* in the face of those who would seek to impede and destroy our dialogue with *Nature*, those we have identified as the enemies of communication. To open ourselves to the other, to *Nature*, as is consistent with any admirable philosophical task, in the ways we have discussed within these chapters, in ways required in order to transform our view, our attuned way of thinking about and responding to *Nature*, is to ultimately *risk* what we have already established through our ever-evolving dialogue. Eco-phenomenology demands that we continually expose ourselves to what is other, to those who challenge the necessary ontological *boundaries* and *thresholds* that we must ethically care for and guard against the attacks and intrusions of outsiders, barbarians, who seek to tear them asunder, in order, in light of our imperative duties, to protect "the victims and the excluded," the needy and the desperate, "the mystics, the mad, the tortured, and the birds and the frogs,"[113] with the goal of fostering, developing, and ultimately, sustaining an ethical attitude toward *Nature* and those with whom we participate in "dialogue." This is ultimately the process of *becoming other in the face of the other*, in the *face of Nature*, or what we have identified as the human and *humane* possibility of *finite transcendence* within our renewed immersed "dialogue" with *Nature*. The analysis in this chapter is intimately relatable to the prima facie duty of *self-improvement*, which is inseparable from the notion of *selfhood* that we explored in relation to *Nature*, and it reminds us that authentic *self-hood* occurs or is made possible *only* through our dealings with *Nature*, and beyond, happens *only* when we *subject* ourselves to, in *subjugation* of, the "reticent" imperatives that *Nature* speaks and vocalizes.

Notes

1 Graham Harman, *Object-Oriented Ontology: A New Theory of Everything* (London: Penguin, 2018), 4.
2 Graham Harman, *Towards Speculative Realism* (Washington: Zero Books, 2010), 6.
3 Ibid., 12.
4 Michael Bonnett, "The Powers that Be: Environmental Education and the Transcendence, *Policy Futures in Education*, 13, no. 1 (2015): 11.
5 Ibid.
6 Michael Bonnett, "Normalizing Catastrophe: Sustainability and Scientism, *Environmental Education Research*, 19, no. 2 (2013): 21.
7 Ibid.
8 Max van Manen, *Researching Lived Experience* (Albany: SUNY Press, 1990), 101.
9 John Gray, *Black Mass: Apocalyptic Religion and the Death of Utopia*, (New York: Farrar, Straus and Giroux, 2007).
10 Martin Heidegger, *Pathmarks*, trans. William McNeill (Bloomington: Indiana University Press, 1998), 107.
11 Bonnett, "The Powers that Be," 13.
12 Ibid., 10.
13 Ibid., 11.
14 Harman, *Object-Oriented Ontology*, 25–33.
15 Ted Toadvine, "The Primacy of Desire and Its Ecological Consequence," in *Eco-Phenomenology: Back to the Earth Itself*, eds. Charles Brown and Ted Toadvine (Albany: SUNY Press, 2003), 142.
16 Bonnett, "The Powers that Be," 10.
17 Michael Bonnett, "Sustainability and Human Being," in *Post-Sustainability and Environmental Education*, eds. B. Jickling and S. Sterling (New York: Palgrave Macmillan, 2017), 6.
18 Charles Brown, "The Real and the Good," in *Eco-Phenomenology: Back to the Earth Itself*, eds. Charles Brown and Ted Toadvine (Albany: SUNY Press, 2003), 16.
19 David Wood, "What Is Eco-Phenomenology?" in *Eco-Phenomenology: Back to the Earth Itself*, eds. Charles Brown and Ted Toadvine (Albany: SUNY Press, 2003), 222.
20 Bonnett, "The Powers that Be," 13.
21 Ibid.
22 Ibid., 16.
23 Ibid., 12.
24 Bonnett, "Sustainability and Human Being," 4.
25 Bonnett, "The Powers that Be," 11.
26 Wood, "What Is Eco-Phenomenology?" 218.
27 Ibid., 220.
28 Ibid.
29 Bonnett, "The Powers that Be," 10.
30 Alphonso Lingis, *The Imperative* (Bloomington: Indiana University Press, 1998), 79.
31 Holmes Rolston III, "Challenges in Environmental Ethics," in *Environmental Philosophy: From Animal Rights to Radical Ecology*, eds. Michael Zimmerman, et al. (Englewood Cliffs: Prentice Hall, 1993), 141.
32 Wood, "What Is Eco-Phenomenology," 221.

33 "Marine Plastic Pollution," www.oceanunite.org.
34 "The 4 O.C.E.A.N. Pillars," https://4ocean.com.
35 Wood, "What Is Eco-Phenomenology?" 213.
36 Ibid.
37 Ibid., 232.
38 Ibid., 213.
39 Ibid., 215.
40 Ibid., 217.
41 Ibid.
42 Ibid.
43 Ibid., 216.
44 Ibid., 217.
45 Ibid., 216.
46 Ibid., 217.
47 Ibid., 216.
48 Bonnett, "Sustainability and Human Being," 3.
49 Martha Nussbaum, *Love's Knowledge* (New York: Oxford University Press, 1990).
50 John Nolt, "The Move from *Good* to *Ought* in Environmental Ethics," *Environmental Ethics*, 28, no. 1 (2006): 371.
51 van Manen, *Researching Lived Experience*, 104.
52 Ibid., 105.
53 Ibid.
54 Randolph C. Wheeler, "Moral Subjectivity and Imperative Subjection" (2006): 2, www.academia.edu/7343439/.
55 Ibid., 4.
56 Ibid., 5.
57 Ibid., 7.
58 Ibid.
59 Wood, "What Is Phenomenology?" 224.
60 Ibid., 215.
61 Michael Bonnett, *Retrieving Nature* (UK: Wiley-Blackwell, 2004), 166.
62 Ibid.
63 Ibid., 167.
64 Bonnett, "Sustainability and Human Being," 5.
65 Bonnett, *Retrieving Nature*, 167.
66 Bonnett, "Sustainability and Human Being," 5.
67 *Lexicon*, 478.
68 Lingis, *The Imperative*, 73.
69 Ibid., 78.
70 Bonnett, *Retrieving Nature*, 167.
71 Ibid., 168.
72 Ibid.
73 Ibid.
74 Wood, "What Is Eco-Phenomenology?" 221.
75 Alphonso Lingis, "Alterity after Infinity: Interview with Alphonso Lingis (2015)," in *Passion in Philosophy: Essays in Honor of Alphonso Lingis*, ed. Randolph C. Wheeler (New York: Lexington Books, 2017), 92.
76 Ibid.
77 Emmanuel Levinas, *Ethics and Infinity: Conversations with Philippe Nemo*, trans. Richard A. Cohen (Pittsburg: Duquesne University Press, 1984), 48.
78 Lingis, *The Imperative*, 74.
79 Alphonso Lingis, *The Community of Those Who Have Nothing in Common* (Bloomington: Indiana University Press, 1994), 45.
80 Lingis, *The Imperative*, 75.

81 Ibid., 78.
82 Ibid., 94.
83 Ibid., 81.
84 Ibid., 72.
85 Ibid., 71.
86 Ibid., 90.
87 Ibid., 74.
88 Ibid., 71.
89 Ibid., 80.
90 Ibid., 90.
91 Ibid., 88.
92 Ibid., 92.
93 Ibid., 98.
94 Ibid., 151.
95 *Webster's New World Dictionary of the American Language* (New York: Simon & Schuster, 1980), 1214.
96 Lingis, *The Imperative*, 91.
97 Ibid., 92.
98 Ibid., 105.
99 Harman, *Towards Speculative Realism*, 20.
100 Melissa Mayntz, "The Importance of Bird Conservation," http://photographyandbirds.wordpress.com/the-importance-of-bird-conservation/.
101 Ibid.
102 Lingis, *The Imperative*, 152.
103 Ibid., 155.
104 Ibid., 154.
105 Ibid., 155.
106 Ibid., 88.
107 Ibid., 73.
108 Ibid., 93.
109 Beth Wellington, "The Myth of Mountaintop Removal Mining," *Guardian*, August 19, 2011, www.theguardian.com.
110 Ibid.
111 Lingis, *The Imperative*, 87.
112 Ibid.
113 Ibid., 87–88.

Bibliography

Angeles, Peter. *Philosophy*. New York: Harper-Collins, 1992.

Aristotle *Topics Books I and VIII*. Translated by Robin Smith. New York: Clarendon Press, 1997.

Arnheim, Rudolf. "The Double-Edged Mind: Intuition and the Intellect." In *The Symbolic Order*, edited by Peter Abbs, 105–119. London: Falmer Press, 1984.

Barrett, Terry. *Why Is It Art? Aesthetics and Criticism of Contemporary Art*. New York: Oxford University Press, 2008.

Best, David. "Feeling and Reason in the Arts: The Rationality of Feeling." In *The Symbolic Order*, edited by Peter Abbs, 70–88. London: Falmer Press, 1989.

Bonnett, Michael. *Retrieving Nature: Education for a Post-Humanistic Age*. United Kingdom: Wiley-Blackwell, 2004.

Bonnett, Michael. "Environmental Education and the Issue of Nature." *Journal of Curriculum Studies*, 39, no. 6 (2007): 707–710.

Bonnett, Michael. *Moral Education and Environmental Concern*. New York: Routledge, 2013.

Bonnett, Michael. "Normalizing the Catastrophe: Sustainability and Scientism." *Environmental Education Research*, 19, no. 1 (2013): 187–197.

Bonnett, Michael. "The Powers that Be: Environmental Education and the Transcendent." *Policy Futures in Education*, 13, no. 1 (2015): 1–15.

Bonnett, Michael. "Sustainability and Human Being." In *Post-Sustainability and Environmental Education*, edited by B. Jickling and S. Sterling, 1–9. New York: Palgrave Macmillan, 2017.

Brown, Charles. "The Real and the Good." In *Eco-Phenomenology: Back to the Earth Itself*, edited by Charles Brown and Ted Toadvine, 3–18. Albany: SUNY Press, 2003.

Brown, Charles and Ted Toadvine. "Eco-Phenomenology: An Introduction." In *Eco-Phenomenology: Back to the Earth Itself*, edited by Charles Brown and Ted Toadvine, vii-xxi. Albany: SUNY Press, 2003.

Chakraborty, Barnini. "Paris Agreement on Climate Change: US Withdraws as Trump Calls it 'Unfair'" (2017). http://foxnews.com.

Critchley, Simon. *Continental Philosophy*. London: Oxford University Press, 2001.

Dahlstrom, Daniel. "Introduction," in *Philosophy of Mind and Phenomenology: Conceptual and Empirical Approaches*. Edited by Daniel Dahlstrom, et al., 1–10. London: Routledge, 2016.

Davenport, Coral. "How Andrew Wheeler, the New Acting EPA Chief Differs from Scott Pruitt" (2018). http://nytimes.com/climate.

Descombes, Vincent. *Modern French Philosophy*. Translated by L. Scott-Fox and J. M. Harding. Cambridge: Cambridge University Press.

Detmer, David. *Phenomenology Explained: From Experience to Insight*. Chicago: Open Court, 2013.

Economides, Louise. "Romanticism and Buddhism: Blake, Heidegger, Buddhism, and Deep Ecology," *Romantic Circles* (2007). www.rc.umd.edu.

Edwards, Paul. "The Critique of the Cosmological Argument." In *Philosophy: Quest for Truth*, edited by Louis Pojman, 74–90. New York: Oxford University Press, 2012.

Embree, Lester. "The Possibility of a Constitutive Phenomenology of the Environment." In *Eco-Phenomenology: Back to the Earth Itself*, edited by Charles Brown and Ted Toadvine, 37–49. Albany: SUNY Press, 2003.

Esbjorn-Hargens, Sean and Michael Zimmerman. *Integral Ecology: Uniting Multiple Perspectives in the Natural World*. Boston: Integral Books, 2009.

Evernden, Neil. *The Social Creation of Nature*. Baltimore: Johns Hopkins Press, 1992.

Gadamer, Hans-Georg. *Truth and Method*, translated by Joel Weinsheimer and Donald Marshall. London: Continuum, 1989.

Gallagher, Shaun. *Phenomenology*. United Kingdom: Palgrave Macmillan, 2012.

Gallagher, Shaun and Dan Zahavi. *The Phenomenological Mind*. London: Routledge, 2008.

Gannon, Megan. "Exxon Valdez 25th Anniversary: 5 Facts About the Historic Spill" (2014). www.livescience.com.

Gillespie, Michael. "The Search for Immediacy and the Problem of the Political Life in Existentialism and Phenomenology." In *A Companion to Phenomenology and Existentialism*, edited by Mark Wrathall and Hubert Dreyfus, 534–545. London: Wiley-Blackwell, 2009.

Gray, John. *Black Mass: Apocalyptic Religion and the Death of Utopia*. New York: Farrar, Straus and Giroux, 2007.

Harman, Graham. "On Vicarious Causation." *Collapse II*, 2, no. 1 (2007): 90–110.

Harman, Graham. *Towards Speculative Realism*. New York: Zero Books, 2009.

Harman, Graham. "Time, Space, Essence, and Eidos: A New Theory of Causation." *The Journal of Natural and Social Philosophy*, 6, no. 1 (2010): 1–17.

Harman, Graham. *The Third Table/Der dritte Tisch*. Ostfildern: Hatje Cantz, 2012.

Harman, Graham. *Immaterialism: Objects and Social Theory*. Cambridge: Polity, 2016.

Harman, Graham. "The Future of Continental Realism: Heidegger's Fourfold." *Chiasma*, 3, no. 1 (2017): 81–98.

Harman, Graham. *Object-Oriented Ontology: A New Theory of Everything*. London: Penguin, 2018.

Harman, Graham. *Speculative Realism: An Introduction*. Cambridge: Polity, 2018.

Heidegger, Martin. *Being and Time*. Translated by R. Macquarrie and E. Robinson. New York: Harper-Collins, 1962.

Heidegger, Martin. *Discourse on Thinking*. Translated by John Anderson and E. Hans Freund. New York: Harper-Perennial Books, 1966.

Heidegger, Martin. *The Question Concerning Technology and Other Essays*. Translated by William Lovitt. New York: Harper Torch Books, 1977.

Heidegger, Martin. "The Origin of the Work of Art." In *Basic Writings*, edited by David F. Krell, 139–212. San Francisco: Harper-Collins, 1979.

Heidegger, Martin. *The History of the Concept of Time: Prolegomena*. Translated by Theodore Kisiel. Bloomington: Indiana University Press, 1992.

Heidegger, Martin. *Holderlin's Hymn "The Ister."* Translated by William McNeill and Julia Davis. Bloomington: Indiana University Press, 1996.

Heidegger, Martin. *Pathmarks*. Translated by William McNeill. Bloomington: Indiana University Press, 1998.

Heidegger, Martin. *Contributions to Philosophy: (From Enowning)*. Translated by. K. Maly and P. Emad. Bloomington: Indiana University Press, 1999.

Heidegger, Martin. *Introduction to Metaphysics*. Translated by Michael Fried and Richard Polt. New Haven: Yale University Press, 2001.

Heidegger, Martin. *The History of Beyng*. Translated by William McNeill and Jeffery Powell. Bloomington: Indiana University Press, 2015.

Holderlin, Friedrich, "Nature." In *Selected Poems and Fragments*. Translated by Michael Hamburger, 17. London: Penguin Classics, 1994.

Horton, Melissa. "What are the Effects of Fracking on the Environment?" (2017). www.investopedia.com.

Husserl, Edmund. *The Crisis of European Sciences and Transcendental Phenomenology: An Introduction to Phenomenological Philosophy*. Translated by David Carr. Evanston: Northwestern University Press, 1970.

Husserl, Edmund. *The Basic Problems of Phenomenology: From the Lectures Winter Semester 1910–1911*. Dordrecht: Springer Press, 2006.

Ihde, Don. *Experimental Phenomenology*. New York: Paragon Books, 1979.

Jardine, David. *To Dwell with a Boundless Heart*. New York: Peter Lang, 1998.

Jaspers, Karl. *Way to Wisdom*. Translated by Ralph Manheim. New Haven: Yale University Press, 1954.

Kaminska, Sonia. "The Alleged Activity of the Active Intellect [Aristotle]: A Wild Goose Chase or a Puzzle to be Solved?" *Zagadnienia Filozoficzne w Nauce*, 54, no. 1, 77–93.

Kant, Immanuel. *Critique of Pure Reason*. Translated by J. M. D. Meiklejohn. New York: St. Martin's Press, 1958.

Kant, Immanuel. *Groundwork of the Metaphysics of Morals*. Translated by Mary Gregor. Cambridge: Cambridge University Press, 1998.

Kohak, Erazin. "'An Understanding Heart: Reason, Value, and Transcendental Phenomenology." In *Eco-Phenomenology: Back to the Earth Itself*, edited by Charles Brown and Ted Toadvine, 19–34. Albany: SUNY Press, 2003.

Langer, Monika. "Nietzsche, Heidegger, and Merleau-Ponty: Some of Their Contributions and Limitations for Environmentalism." In *Eco-Phenomenology: Back to the Earth Itself*, edited by Charles Brown and Ted Toadvine, 103–120. Albany: SUNY Press, 2003.

Leather, Amy. "Hopelessly Devoted to Fossil Fuels." *Socialist Review* (2017). http://socialistreview.org.uk/420/hopelessly-devoted-fossil-fuels.

Levinas, Emmanuel. *Totality and Infinity*. Translated by Alphonso Lingis. The Hague: Martinus Nijhoff, 1961.

Levinas, Emmanuel. *Ethics and Infinity: Conversations with Philippe Nemo*. Translated by Richard A. Cohen. Pittsburg: Duquesne University Press, 1984.

Liddell and Scott. A *Lexicon: Liddell & Scott's Greek–English Lexicon*. Mansfield Centre: Martino Publishing, 2015.

Lingis, Alphonso. *Phenomenological Explanations*. Boston: Martinus Nijhoff Publishers, 1989.

Lingis, Alphonso. *The Community of Those Who Have Nothing in Common*. Bloomington: Indiana University Press, 1994.

Lingis, Alphonso. *Foreign Bodies*. London: Routledge, 1994.

Lingis, Alphonso. *The Imperative*. Bloomington: Indiana University Press, 1998.

Lingis, Alphonso. "Alterity after Infinity: Interview with Alphonso Lingis (2015)." In *Passion in Philosophy: Essays in Honor of Alphonso Lingis*, edited by Randolph C. Wheeler, 91–104. New York: Lexington Books, 2017.

Llewelyn, John. "Prolegomena to Any Future Phenomenological Ecology." In *Eco-Phenomenology: Back to the Earth Itself*, edited by Charles Brown and Ted Toadvine. 51–72. Albany: SUNY Press, 2003.

Magrini, James. "Thinking Through Nihilism: The Reclamation of Embodied Thought for Enhancing Cultural Practices." *Analysis and Metaphysics*, 12, no. 1 (2013): 133–153.

Magrini, James and Elias Schwieler. *Reading Heidegger After the "Turn": At the Limits of Metaphysics*. New York: Routledge, 2017.

Marcel, Gabriel. *The Mystery of Being (Vol. I)*. Translated by G. S. Fraser. New York: Gateway, 1959.

Marietta, Don. "Back to Earth with Reflection and Ecology." In *Eco-Phenomenology: Back to the Earth Itself*, edited by Charles Brown and Ted Toadvine, 121–138. Albany: SUNY Press, 2003.

Matthews, Eric. *The Philosophy of Merleau-Ponty*. Ithaca: McGill-Queen's University Press, 2002.

Mayntz, Melissa. "The Importance of Bird Conservation" (2017). http://photographyandbirds.wordpress.com/the-importance-of-bird-conservation.

Merleau-Ponty, Maurice. *Prose of the World*. Translated by John O'Neill. Evanston: Northwestern University Press, 1964.

Merleau-Ponty, Maurice. *Phenomenology of Perception*. Translated by C. Smith. London: Routledge, 1989.

Naess, Arne. "The Shallow and the Deep, Long-Range Ecology Movement: A Summary." *Inquiry*, 16, no. 1 (1986): 95–100.

Naess, Arne. "The Deep Ecological Movement: Some Philosophical Aspects." In *Environmental Philosophy: From Animal Rights to Radical Ecology*, edited by Michael Zimmerman, et al., 193–211. Englewood Cliffs: Prentice Hall, 1993.

Nolt, John. "The Move from *Good* to *Ought* in Environmental Ethics." *Environmental Ethics*, 28, no. 1 (2006): 355–374.

Nussbaum, Martha. *Love's Knowledge*. New York: Oxford University Press, 1990.

Palast, Greg. "Court Rewards Exxon for Valdez Oil Spill" (2008). www.gregpalst.com.

Pattison, George. *Later Heidegger*. London: Routledge, 2000.

Peters, F. E. *Greek Philosophical Terms: A Historical Lexicon*. New York: New York University Press, 1967.

Pojman, Louis. *Ethics: Discovering Right and Wrong*. Belmont, CA: Wadsworth Press, 2009.

Polt, Richard. *The Emergency of Being: On Heidegger's Contributions to Philosophy*. Ithaca: Cornell University Press, 2006.

Popovich, Nadia and Tatiana Schlossberg. "23 Environmental Rules Rolled Back in Trump's First 100 Days" (2017). www.nytimes.com.

Pufendorf, Samuel. *On the Duty of Man and Citizen According to Natural Law*. Cambridge: Cambridge University Press, 1991.

Putnam, Hilary. *The Many Faces of Realism*. LaSalle: Open Court, 1987.

Read, Daniel. "Exxon Valdez: Corporate Greed and Environmental Catastrophe" (2009). www.marxism.com.

Reid, Louis Arnaud. "The Arts Within a Plural Concept of Knowledge." In *The Symbolic Order*, edited by Peter Abbs, 12–21. London: Falmer Press, 1989.

Rilke, Rainer Maria. *Letters to a Young Poet*. Translated by M. D. Herter Norton. New York: W. W. Norton & Company, 2004.

Rolston III, Holmes. *Environmental Ethics: Duties to and Values in the Natural World*. Philadelphia: Temple University Press, 1981.

Rolston III, Holmes. "Challenges in Environmental Ethics." In *Environmental Philosophy: From Animal Rights to Radical Ecology*, edited by Michael Zimmerman, et al., 135–157. Englewood Cliffs: Prentice Hall, 1993.

Ross, W. D. *The Right and the Good*. New York: Oxford University Press, 1961.

Sartre, Jean-Paul. *Jean-Paul Sartre: Essays in Existentialism*. Translated by Wade Baskin. New York: Citadel Press, 1995.

Scully, Vincent. *The Earth the Temple and the Gods*. New Haven: Yale University Press, 1997.

Sprintzen, David. *Camus: A Critical Examination*. Philadelphia: Temple University Press, 1998.

Steinmann, Michael. "Phenomenological Perspectivism," *Heidegger Circle Archive*. 45th Annual Meeting (2004).

Tallis, Raymond. *Aping Mankind: Neuromania, Darwinitis, and Misrepresentation of Humanity*. Durham: Acumen, 2011.

Thomson, Iain. "Ontology and Ethics at the Intersection of Phenomenology and Environmental Philosophy." *Inquiry*, 47, no. 1 (2004): 380–412.

Thomson, Iain. "Environmental Philosophy." In *A Companion to Phenomenology and Existentialism*, edited by Mark Wrathall and Hubert Dreyfus, 445–463. London: Wiley-Blackwell, 2009.

Toadvine, Ted. "The Primacy of Desire in Its Ecological Consequences." In *Eco-Phenomenology: Back to the Earth Itself*, edited by Charles Brown and Ted Toadvine, 139–154. Albany: SUNY Press, 2003.

Turner, Kerry. "Sustainability, Resource Conservation and Pollution Control: An Overview." In *Sustainable Environmental Management*, edited by Kerry Turner, 10–23. London: Belhaven Press, 1998.

van Manen, Max. *Researching Lived Experience*. Albany: SUNY Press, 1990.

Wellington, Beth. "The Myth of Mountaintop Removal Mining" (2014). www.theguardian.com.

Wheeler, Randolph C. "Moral Subjectivity and Imperative Subjection" (2006). http://Academia.edu/7343439/.

Wheeler, Randolph C. "Imperative Innovations." In *Passion in Philosophy: Essays in Honor of Alphonso Lingis*, edited by Randolph C. Wheeler, 119–133. New York: Lexington Books, 2017.

Williamson, David. "Exxon Valdez Oil Spill Effects Lasting far Longer than Expected." *UNC News University of North Carolina at Chapel Hill* (2013). www.unc.edu.

Wood, David. "What Is Eco-Phenomenology?" In *Eco-Phenomenology: Back to the Earth Itself*, edited by Charles Brown and Ted Toadvine, 211–234. Albany: SUNY Press, 2003.

Wrathall, Mark and Hubert Dreyfus. "A Brief Introduction to Phenomenology and Existentialism." In *A Companion to Phenomenology and Existentialism*, edited by Mark Wrathall and Hubert Dreyfus, 5–11. Malden: Wiley-Blackwell, 2008.

Zahavi, Dan. "The Phenomenological Tradition." In *Routledge Companion to Twentieth Century Philosophy*, edited by D. Moran, 8–21. London: Routledge, 2008.

Zahavi, Dan. *Husserl's Phenomenology*. Stanford: Stanford University Press, 2008.

Zimmerman, Michael. "Heidegger, Buddhism, and Deep Ecology." In *The Cambridge Companion to Heidegger*, edited by Charles Guignon, 293–321. Cambridge: Cambridge University Press, 1993.

Zimmerman, Michael. *Heidegger's Confrontation with Modernity: Technology, Politics, Art*. Bloomington: Indiana University Press, 1990.

Index

For Product Safety Concerns and Information please contact our EU representative GPSR@taylorandfrancis.com
Taylor & Francis Verlag GmbH, Kaufingerstraße 24, 80331 München, Germany

www.ingramcontent.com/pod-product-compliance
Ingram Content Group UK Ltd.
Pitfield, Milton Keynes, MK11 3LW, UK
UKHW011011180926
475701UK00004B/687

* 9 7 8 0 3 6 7 7 2 8 2 2 9 *